Courtroom Testimony for Emergency Responders

COURTROOM
TESTIMONY
for
EMERGENCY
RESPONDERS

CLIFF MUNSON

Fire Engineering

Copyright© 2013 by
PennWell Corporation
1421 South Sheridan Road
Tulsa, Oklahoma 74112–6600 USA

800.752.9764
+1.918.831.9421
sales@pennwell.com
www.FireEngineeringBooks.com
www.pennwellbooks.com
www.pennwell.com

Marketing Manager: Amanda Alvarez
National Account Manager: Cindy J. Huse

Director: Mary McGee
Managing Editor: Marla Patterson
Production Manager: Sheila Brock
Production Editor: Tony Quinn
Cover Designer: Charles Thomas

Library of Congress Cataloging-in-Publication Data

Munson, Cliff.
  Courtroom testimony for emergency responders / Cliff Munson.
     pages cm
  Includes bibliographical references and index.
  ISBN 978-1-59370-323-3
1. Evidence, Expert--United States. 2. Arson investigation--United States. 3. Fire investigation--United States. 4. Fire fighters--Legal status, laws, etc.--United States. 5. First responders--Legal status, laws, etc.--United States. I. Title.
  KF9674.M86 2013
  345.73'067--dc23

                                    2013031118

Printed in the United States of America
    1 2 3 4 5   17 16 15 14 13

For Jackie, Kristina, Erin, Taylor, Jon, and Lola, without whom none of this would have happened.

Contents

Foreword

The fire service in the United States, like all components of public safety, has undergone tremendous changes ever since residents of a community determined there was a need for such service. In the fire service particularly, there has been a gradual shift from all volunteer to volunteer/paid departments and then finally an expansion to fully paid departments. And just like the changes worked over the years throughout the fire service, I was able to witness these same changes within my own department in Long Beach, California. Among the many changes, and probably most paramount, was the shift from a strictly "fire service" mentality to one of comprehensive emergency/ hazard management. This setting, where EMS, hazmat, and other types of responses became the statistical majority, was one to which both Captain Cliff Munson and I were introduced at the start of our careers. Each department in its own way began a shift toward response expansion along with much stricter resource allocation and management. Fire service personnel were now forced to become experts in more than just the field of fire suppression—they now became specialists/experts in EMS, hazmat, and arson investigation.

Captain Munson was one of those far-sighted individuals who realized that his value to his department as well as his own sense of fulfillment would be contingent upon his seeking and attaining additional expertise during his entire time on the department, and he went about this aggressively. He actively sought and obtained additional education beyond a degree in fire science, obtaining a bachelor's degree as well. Whenever the situation presented itself, he would attend classes offered by the state fire marshal. At the earliest opportunity he applied through department channels and was accepted to attend Los Angeles County Paramedic School. He successfully completed an intensive six-month course there and performed in the capacity of a state certified paramedic at some of Southern California's busiest stations. Finally, in 1990 he attained the rank of company officer—a captain with Long Beach (CA) Fire Department.

But even here, at a point most individuals seek to attain and finish out their careers, Captain Munson sought yet another challenge—that of captain/head of investigations with the department's Arson Unit. Prior to his arrival, the Arson Unit had consisted of two full-time and one or two part-time investigators. While the unit had an outstanding conviction rate, it was geared mainly toward fires resulting from arson. Realizing a need for additional investigative ability, Captain Munson applied to be the head of the Arson Unit and was accepted. At this point, he again sought to excel and improve an already well-functioning environment. Because of the variety of incidents that were now being investigated under his command, the name of the unit was changed to Long Beach Fire Department Investigation Unit. To meet the burgeoning number of hazmat incidents to which units were responding, a full-time Long Beach police sergeant was put on permanent assignment with the Investigation Unit to handle the volume of responses and provide input on case handling. An additional investigator's position was added to help with both arson and hazmat incidents.

During this time, as stated above, resource allocation became a priority in a public environment with decreasing revenue. Captain Munson saw here an ability to help minimize the financial impact on his department and was able to make the Investigation Unit self-sustaining within a short time, thereby offsetting costs of personnel and equipment. He also sought training for the department as a whole, along with extensive training and equipment for those personnel in the Investigation Unit.

As you go through this text, you will see where he has placed emphasis upon those things most young members of the fire service pay little attention to at the start of their careers. These include what happens after the fire is extinguished, things such as basic investigation procedures, background on the Fourth and Fifth Amendments, and courtroom protocol. He also stresses the importance to citizens of thorough investigative procedures, particularly to family members of the dead or injured, thereby bringing closure to them for what has been undoubtedly a tragic incident.

One incident in particular where I was able to assist and see first-hand the competence and dedication Captain Munson brought to all his investigations occurred in early 1997 and involved arson with great bodily injury to a victim along with extensive damage to a dwelling. I was a battalion chief at the time, and one of my assignments was public information officer. Because there were a number of Spanish-speaking witnesses, I was asked to respond and assist with witness interrogation. It gave me an opportunity to see first-hand how far the Investigation Unit had come in a short time. A variety of techniques were used, including an arson dog to detect the presence of hydrocarbons on the suspect, along with various imaging and recording devices. There was also an effective liaison with the Long

Beach Police Department when it came time to transport the suspect to custody and perform additional interviews. As a result, the suspect was convicted of great bodily injury resulting from arson along with attempted murder and received a lengthy prison sentence. Because the victim was mistaken for someone else by the suspect, it was wonderful for the victim and his family to have the closure brought about by that prison sentence. This was not the only incident we worked on; often I would provide information to the local news media regarding significant fires in the form of the "always credible" information given to me by members of the Investigation Unit. As a result of Captain Munson's efforts, it became well known that the Long Beach Fire Department was effective in not only managing emergency incidents but handling the particulars surrounding the investigation and prosecution of those incidents as well.

As you go through the text, I ask you to do what most young firefighters don't—that is to see yourself in the role of an arson investigator. Imagine what you are going to do *after* the fire, and allow yourself the latitude to expand your educational pursuits toward that goal. As you can see, if done properly, it is an extremely rewarding career.

Reed Bingham
Battalion Chief (Ret.)
Long Beach Fire Department
Long Beach, California

Acknowledgments

Special thanks to:
Bill Klein
Chemeketa Community College
Debbie Bates
Erin Gorham
Honorable Courtland Geyer
Jacqueline G. Munson
Jon Munson
Larry Feller
Long Beach (California) Fire Department
Marion County (Oregon) Circuit Court
Reed Bingham
Taylor Munson
2012 Chemeketa Storm Baseball Team

In memory of Deputy Tyler Chapman.

Introduction

There is no substitute for experience when it comes to courtroom testimony. Unfortunately, experience is something you get immediately after you needed it. Experiences can be both good and bad, and most of the time, courtroom experiences turn out to be bad for at least half the people participating; sometimes for everyone. I have learned that careful preparation and knowing what to expect from your "day in court" can cause this experience-gathering journey to be a much more palatable one.

My book is not going to make you an expert witness. My book is not going answer every question you might have about testifying in a courtroom. What it will do is give you some insight into what you can expect when you enter a courtroom as a witness or investigating officer and what is expected of you. When you finish this book, you should have a good understanding of all the players at a trial, their respective roles, and where you fit in. These insights will help you to prepare and present yourself as an informed and skillful investigator at trial.

I am not an attorney, and this book is not going to give you legal advice. I am sure that many attorneys will not agree with some of the statements and observations here, but getting a group of attorneys to agree with me is not my goal. I am a firefighter who, after 17 years on the line, decided to venture into the field of fire and arson investigation for the last 15 years of my career. I have been in the fire service for more than 32 years, and I have been exposed to most of what there is to experience. Those experiences are the basis of this book. I did not sit down and read several other books and then consolidate the experiences of others into a book. You will note a lack of many footnotes and any bibliography citing other people's books as the basis for mine. This book is completely drawn on my daily experiences in court and dealing with the legal system as an arson investigator.

In 1994 I became the commanding officer of the Investigation Unit for the Long Beach Fire Department in Southern California. Long Beach is a sizable, ISO Class I Fire Department that conducts fire and arson investigations the way they should be done. The turnover in the arson unit is very small, and the men and women who work in it are well educated, experienced, and dedicated to their jobs. The management of the Long Beach Fire Department supports the unit wholeheartedly, even when the job causes some unpleasant interaction with local residents, as it often does. I was lucky to work in Long Beach because they did do the job correctly, and not that many fire departments do. When you're an arson investigator in Long Beach, you're an arson investigator and nothing else. You are in or you are out. Being a good fire investigator or arson investigator takes complete dedication to staying current on the law, following up on investigations, and filing reports in a timely manner. It is very hard to do this correctly when you are trying to split your time between inspections, puppet shows, and other duties.

All investigators in the Long Beach Fire Department Investigation Unit are required to complete a police academy and are sworn peace officers. They carry firearms, and they make their own arrests, write and file their own reports with the Los Angeles County district attorney, and serve as investigating officers for the DA's office. They were all certified firefighters with years of experience in the suppression ranks before becoming members of the Investigations Unit. It is a very competitive process to become a member of the unit and if you are chosen to serve, you are well qualified to do so. You are also expected to devote that portion of your career to fire and arson investigation and the duties that come with it.

During my time in Long Beach, I investigated hundreds of fires and several other crimes involving fire, explosives, or the attempt and threat to use those means to harm others. I participated in 135 felony arrests and many lesser arrests or citations for various misdemeanor crimes. I have testified in court nearly 100 times as both an expert witness and a factual witness in civil cases and criminal trials. I have testified in California and Oregon, and I have testified at every level of crime from misdemeanor reckless use of fireworks all the way up to murder by arson. I not only served as the commanding officer of the Investigation Unit for the Long Beach Fire Department, but also as a deputy state fire marshal in Oregon, and finally as the deputy fire marshal and chief investigator for a small fire department west of Portland, Oregon. I have spent a lot of time in court, and as a result of that experience, I have accumulated a lot of firsthand knowledge. I learned that many investigators and police officers enter the courtroom unprepared to do battle with the defense attorney and don't understand the legal system. I want to help you prepare for that battle.

As you go through the chapters of this book, you will see topics that you have heard of before. NFPA 921 and *Michigan v. Tyler* will be among them. This book explores these topics not from the standpoint of legal decisions, but regarding what you need to know to when you raise your right hand. Ideally, what you learn from this book will keep you from making mistakes that can make your life miserable. Everything here stems from my personal experience, and I have written it in a firefighter-to-firefighter or investigator-to-investigator manner. In other words, it's plain talk that cuts through the jargon and chaff. When appropriate, I have used real-life examples of cases I was involved with to illustrate the points I am making, and I have tried to do it with some humor.

Let me end this introduction by thanking you for reading this book. I hope in some small way it will help you in the near future. My personal email address is in the back of this book, and I invite you to contact me with comments or questions, or to talk about interesting cases.

FESHE Objectives Correlation

This text is written to meet the course outcomes for the Political and Legal Foundations for Fire Protection curriculum established in the Fire and Emergency Services Higher Education (FESHE) Model Curriculum. Outcomes are met in the following chapters:

Objective	Chapter(s)
1. Describe the differences between different types and branches of law.	3, 4, 5
2. Identify classes of actions such as tort, contract, and property.	3, 4, 5, 8
3. Explain the system of law and its functions.	1, 2, 3, 5, 7
4. Explain the importance to the fire service of the due process clause of the Fourteenth Amendment.	3
5. Define criminal and administrative warrants.	6
6. Describe the circumstances requiring warrants and exceptions.	6
7. Define sovereign immunity and Good Samaritan protection as they relate to the fire service.	1, 9

1

Where Do I Start?

Learning Objectives

Upon completion of this chapter, you should be able to:

- Explain the importance of being well trained and well certified when preparing to appear in court.
- Describe some of the resources available to achieve this training.
- Define the intent of the National Fire Protection Association (NFPA).
- Describe the different ways in which fire investigation is conducted from jurisdiction to jurisdiction.
- Describe some methods for gaining experience as a fire investigator.

Case Study

The young arson investigator was sitting in the hallway of the courtroom waiting to speak with the prosecutor concerning the case he had spent the last three weeks working on and the arrest that resulted from his hard work. He had joined his fire department's arson detail about two months ago, and this was his first chance to appear in court. He had participated in a mock trial during a weekend fire investigation class, which he had attended six months ago, but the opportunity to attend a real trial had never presented itself.

After a while, a rather stern-looking individual in a business suit approached him and asked if he was the investigator who had filed the criminal case for arson. The investigator answered in the affirmative, but could tell by the look on the face of the prosecutor that something was not right. The investigator asked if everything was all right, and the prosecutor responded by saying the case appeared to be mostly circumstantial, and they would be relying heavily on the investigator's training and expertise to make the case. The prosecutor went on to tell the young investigator that often these types of cases turn on the investigator winning the confidence of the jury and the judge in the trial to declaring the investigator to be an expert in his field through experience, education, and certification.

At this point the prosecutor asked the investigator to go over his experience in the field of arson investigation and to briefly list the professional certifications he held. The young investigator then told the prosecutor that he had attended a weekend fire investigation class at a neighboring fire department about six months ago, but he had not had time to attend any other classes or gain any certifications at all. He went on to explain that his fire department was pressed for money and that the chief wanted him to take over the fire investigation duties because he needed someone. Because he was interested, the chief chose him.

At this point the prosecutor heaved a sigh, looked at the young investigator, and said, "We have a problem!"

Introduction

Fire investigators are among the most dedicated of fire department employees and often spend their own money, time, and a great deal of energy to become the professionals they need to be. It's important for potential fire investigators to understand that fire investigation often leads to uncovering criminal activity, which entails legal issues. Victims of arson often depend upon the investigator to make a strong enough case for them to have some type of justice for their suffering. Thus, the investigator must often make several appearances in a court of law.

Prosecutors are attorneys who are well versed in the law, but they are not actually fire investigators. As a rule, they depend heavily on the expertise and credentials of fire investigators to help them win convictions in important criminal cases. When prosecutors do not possess the bona fides necessary to win a conviction, arsonists are sometimes allowed to walk out of a courtroom without paying their debt to society.

Most jurists are good people who want to do the right thing, but they cannot in good conscience convict anyone of a crime unless the evidence and the expert testimony are there. It is up to the fire or arson investigator to provide them with that evidence.

Gaining Credentials

If you wish to be taken seriously in a court of law, you must develop a set of credentials that will stand up to the scrutiny of a judge and the skepticism of a jury. You are never going to be perceived as a credible witness in court unless you take the time to develop a resume that includes important certifications, meaningful formal education, and, of course, experience.

During your career as a fire or arson investigator, your credentials and level of expertise are going to be challenged during every arson case you testify in. In many cases, discrediting you as a witness is going to be the only avenue open to the defense, and they will, without shame or hesitation, proceed right down that avenue.

If it becomes necessary for you to qualify as an expert witness in an arson case, you can count on your credentials being challenged by the defense attorney. Her job is to provide the defendant with the best possible defense, and challenging your qualifications to testify as an expert witness is part of her job (fig. 1–1). You should be prepared to face this. It's important to make the distinction between expert witness and **factual witness**, which will be discussed later in this book (fig. 1–2). For now, assume that when the word *witness* is used, we are talking about you, the fire investigator, arson investigator, or first responder in the case, and you are testifying about an investigation you conducted or occurrences you witnessed at the scene of an emergency.

Fig. 1–1. Firefighters are often called upon to function as expert witnesses. This requires a great deal of specialized training, experience, and knowledge. Once qualified as an expert, your opinion can be rendered under oath.

Fig. 1–2. Civilians and nonexpert witnesses are usually limited to testifying to what they heard, saw, or did. Their opinions in cases are not allowed to be stated while on the stand.

Fire Investigation Task Forces

One of the challenges for the fire investigator is the lack of any clear-cut guidelines about how much training is necessary, how many certificates, and which ones will make you credible in a court of law. The level of training and certification varies wildly from state to state, and many individuals who view themselves as fire investigators are in fact far from it. One practice that is common in rural areas is the formation of a **fire investigation task force,** called a FIT team. It should be clearly stated that fire investigation task forces do exist that are populated by highly trained and highly documented investigators who provide fire investigation services at major fires or large life-loss fires. These task forces include highly respected members of organizations such as the Federal Bureau of Investigation (FBI) or Bureau of Alcohol, Tobacco, Firearms and Explosives (ATF). However, this book is not concerned with these types of task forces.

There are some cases in which FIT teams are composed of large groups of people loosely associated with a fire department, who form groups to investigate fires. These groups range anywhere from 10 to 50 people, and their goal is to show up at fires and conduct group investigations. While most of the members of these FIT teams are well intentioned and do a good job, there are varying levels of training and experience in these large groups. Sometimes that leads to members placed in positions within the group that are beyond their level of training, expertise, and experience. Remember that during a courtroom appearance, the fire investigation task force member will the need to defend his or her credentials and training. Very important criminal and civil cases sometimes turn on the qualifications of the investigator, and oftentimes those qualifications are just not sufficient.

National Fire Protection Association Standard 1033

NFPA 1033: *Standard for Professional Qualifications for Fire Investigation* is a fine document as far as it goes. It sets the minimum qualifications for a fire investigator, and it outlines such important topics as age requirements and background investigations for potential fire investigators, and it even mentions that a fire investigator must be able to testify in court and to be credible while doing so. Note that NFPA 1033 does not definitively state how many hours of training and what type of training are needed to become a competent fire investigator.

Though NFPA 1033 does not specify training hours or required certification for the fire investigator, the answer is simple; the learning never stops. There is no fire investigator alive who is so smart, so well trained, and so completely up on all current information that there is not another thing he or she needs to learn. The time will come when you will be on the witness stand, and the prosecutor will run through your resume of training, certifications, and experience for the judge in the trial to determine if you can be called an expert. When that day comes, it is incumbent upon you to have taken advantage of every opportunity to advance your knowledge and skill in the field of fire investigation. The job of fire or arson investigator requires constant training and an awareness of what is occurring in the field at all times.

Fire Investigator Versus Arson Investigator

Let us take a moment and define the difference between an **arson investigator** and a **fire investigator**. The differences are significant, so it is important that you know which you are. These two titles or job descriptions are by no means interchangeable for the most part. There are many people in the United States who have the training and authority to fill both positions simultaneously, but that is usually not the case. An apt comparison might be that of baseball and softball. Several features of these sports are quite similar, such as the layout of the field and many of the rules. In both baseball and softball, the batter is out after three strikes, there are three outs in each inning, and nine defensive players are on the field at any one time. But we all know these two games are radically different when it comes to the details. A similar comparison can be drawn between fire investigator and arson investigator. For the purposes of this book, a person with either job description has an substantial chance of testifying in a criminal or civil trial.

So, let's start by defining the two jobs. A fire investigator is a person who is trained, experienced, and certified to some extent in the practice of examining a burned building after the fact and attempting to discover where and how the fire started. Like it or not, the International Fire Code places the responsibility for fire investigation squarely on the shoulders of the fire department. Chapter 1 of the International Fire Code clearly states this and leaves no doubt that the authority and responsibility for fire investigation lie with the fire code official. The code does not, however, elaborate to any degree what type of an investigation should be accomplished and to what extent the fire department is responsible. In some departments, the engine or truck company officer bears the full responsibility for conducting an investigation. In other departments, the duty is part of the job description of the fire inspector in the fire prevention bureau. In departments that do not have a fire prevention bureau, this duty may be kicked to another agency such as the State Fire Marshal's Office. The levels of training and expertise vary greatly from agency to agency. (This will be explored in greater detail later in this chapter.) Most fire investigators in the public sector are employed by fire departments or agencies similar to fire departments. This is not a hard-and-fast rule, and there are probably exceptions, but for the most part this is the case. Most fire investigators do not enjoy the police powers of a law enforcement officer, and when they discover they are investigating a criminal act, they will generally call for law enforcement at that time and either work with the law enforcement officers or turn the investigation over to them completely.

An arson investigator is an individual who possesses all of these skills and certifications of a fire investigator and, in addition, the training, certification, and authority of a law enforcement officer. Note that just because an individual is a certified law enforcement officer does not necessarily mean he or she works for a law enforcement agency. Most municipalities take the single job of arson investigator and divide it up among at least two agencies and sometimes three or four. This presents a challenge in maintaining control over the investigation and potentially the outcome of the trial, if the incident is deemed a criminal act. There are many other fire departments with good arson investigators who are able to follow the case from beginning to end. In these cities the arson investigators conduct the investigation of the fire, develop both suspects and witnesses in the case of criminal acts, make the arrest, interview the suspect, and then present the case to the district attorney for prosecution. At trial, the same investigator serves as an aide to the prosecutor and provides all of the expertise needed to understand how the fire was started and who started it. Armed with this knowledge and assistance, the prosecutor can make a meaningful presentation to a jury that will ideally result in the conviction of an arsonist. Many of the qualities that would make a person a good firefighter will lend themselves to making that same individual a good law enforcement officer. Many agencies find that sending a good fire investigator to a police academy to teach him or her to be a law enforcement officer is the best and most economical way to accomplish the job. It eliminates duplication of effort, requires fewer personnel to accomplish the same mission, and yields a more seamless investigation.

While most people would state that having a fire department arson investigator sounds like the best approach to investigating fires, many departments avoid it for fear of looking too much like a law enforcement agency. If you are going to be a fire or an arson investigator, you need to come to grips with the attitude in your jurisdiction and deal with it. Regardless of which approach your agency chooses to provide fire investigation services to the residents of your community, you have an excellent chance of ending up in court, so it would make sense to have the best trained and empowered investigator as part of the department.

Getting the Necessary Training

If you are going to be a fire investigator, you should take the time to train yourself to be one. This means that you are going to have to take some initiative. In an era when fire department budgets are stretched, it may be necessary for you to receive training at your own expense. This is particularly true for those who are not yet employed by a fire department, but are training themselves for potential employment. There is a pervasive attitude in some fire departments, or at least among their members, that if a particular type of training is needed, the fire department will supply it to me. This is definitely not true of fire investigation. The members of the fire department who end up with investigative duties are usually highly motivated individuals who sought training on their own with the intent of attaining a level where they could serve their department in that capacity. Most chief officers would like to put a motivated firefighter in the position of fire investigator.

Very good training is available all across the United States, and it's available to anyone willing to spend the time and energy to attend the classes. Foremost among this training is the National Fire Academy (NFA) in Emmetsburg, Maryland. The NFA provides a two-week course in fire investigation. The course includes both classroom instruction and hand-on practice in burn rooms. You must first get the approval and endorsement of your fire department to attend the training, and the application process should be followed to the letter to make sure you will be taking the correct course and in the correct time frame to be considered. This is a great opportunity for new investigators to start and a great place for experienced investigators to brush up on their skills.

Many colleges and universities provide resident and online courses in fire investigation as well. There are semester-long courses that deal exclusively with fire investigation, and many of these courses include several live burn exercises as part of their curriculum. While these courses in themselves will not make you an expert in fire investigation, they will give you a good beginning in the process. Community colleges are a huge untapped resource available to individuals wishing to become investigators.

The **International Association of Arson Investigators (IAAI)** is an international organization that conducts seminars in virtually every state in the union, generally year-round. The IAAI is without doubt one of the finest organizations involved in the training of fire and arson investigators (fig. 1–3). It has a very user-friendly website that offers extensive training opportunities to those who are willing to invest the time and effort in its outstanding classes. There are currently over 40 training modules available, free of charge. Its Certified Fire Investigator (CFI) trainer website contains a wealth of training available to the devoted student. Many state agencies and the NFA are using these courses as prerequisites for applicants to their fire investigation programs.

These seminars are available in both an online and a face-to-face format to provide opportunities for investigators to listen to experts in their field and to interact with investigators from other parts of their state and the country. This interaction sometimes leads to new and interesting ways to improve a fire department's investigation procedures. There is no place in the fire investigation field for an attitude of "we have always done it this way." Court decisions and technology will constantly change the manner in which we investigate fires, and we must stay abreast with these changes. The IAAI is a cutting-edge organization in this field.

The IAAI offers its IAAI-CFI certification, which is widely recognized and well respected. This is exactly the type of certification that judges look for when deciding whether or not to declare a witness to be an expert in the field of fire investigation (fig. 1–4).

Fig. 1–3. The International Association of Arson Investigators (IAAI) is an outstanding worldwide organization that offers training to arson and fire investigators.

Fig. 1–4. New arson and fire investigators must seek training and education on their own. Here the IAAI provides a class in structure fire investigation.

Other Seminars

Many local fire departments offer weekend seminars on specific topics such as electrical fires or incidents involving engine compartments of vehicles. This type of specialized training is valuable and a welcome addition to anyone's training resume. The training is there if you take the time to attend (fig. 1–5).

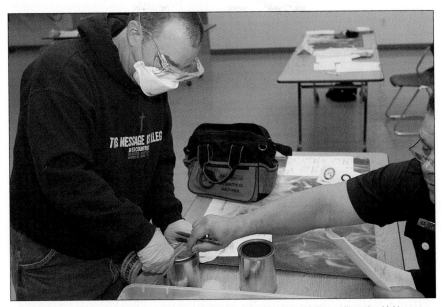

Fig. 1–5. Training should not be limited to the cause and origin of fires. Here the IAAI provides training in forensic evidence collection.

Gaining Experience

Most fire investigators will tell you that there simply is no substitute for experience. It is a great advantage to have been a firefighter before venturing into the field of fire investigation. Such fire investigators benefit greatly from their years of experience in fire suppression. Many fire investigators do not begin their careers with the desire to be an investigator, but as rookie firefighters, they followed the fire investigators around burned-out buildings as they performed their duty. These investigators relate that they were fascinated with how the investigators could look at what appeared to be nothing more than the charred rubble of a building and spot signs and indicators about where the fire might have originated. Experience can be gained by following investigators through a building from the point of least burned to the point of most burned, watching them do a complete walk around the exterior of the building, and finally observing them search trash cans in a two-block area when they suspect arson.

The more chances young firefighters have to watch investigations being conducted, the more sense the investigators' tactics make. Soon, suppression firefighters will begin to anticipate how the investigators will conduct their investigations and where the telltale signs might lead them. They actually begin to see the **V patterns** the investigators talk about and begin to understand their methods and objectives. These young firefighters are gaining experience, and they don't even know it. In many cases, having the opportunity to observe fire investigations being performed for many years becomes the motivation for young firefighters to become fire and arson investigators.

This doesn't mean that you cannot be a good fire or arson investigator unless you were first a line firefighter. Although your path may be more difficult, with enough effort and dedication, you can achieve the goal of becoming a good fire or arson investigator. You will need to find a way to gain experience and to work some actual fires before any credibility will attach to your career, but that can be done in several ways.

Internships

Many fire departments in the United States will allow an internship within a specific part of the fire department. By definition, an internship involves being exposed to experienced people performing the job to which the intern aspires. In return, the intern performs various functions for the department, which not only enriches the intern but also benefits the fire department providing the experience.

For example, it is a good idea for anyone who wants to be a fire investigator to spend as much time as possible with experienced investigators, watching them work fire scenes. In return for this experience, the intern usually assists the investigators around the office with tasks such as filing or performing maintenance on the equipment used during the investigation. Questions are frequently asked of the investigators, and what sometimes transpires is a one-on-one lesson on some aspect of fire investigation.

As an intern, you could reasonably expect to learn about methods of investigation and **interrogation**, how to file a criminal case with the **district attorney,** and many other valuable lessons that would not normally be available to a nonmember of the fire department. While most people probably believe they have better things to do with their time than hang around a fire department, this type of dedication and initiative do not usually go unnoticed.

Policies for **ride-alongs** and internships vary from department to department, as do investigation practices. It would be prudent to check with the fire departments you would like to ride with and find out if they offer such programs and what you could expect to experience during the internship or the ride-along.

Community Colleges and Universities

Community colleges that have fire science or fire technology programs usually offer courses in fire investigation. These courses are normally a good way to start laying the foundation to become an investigator (fig. 1–6). Many of these courses have a lab component that requires students to investigate actual fire scenarios, interview witnesses and suspects, and then present their findings in report form that will be evaluated as if they were presenting the case for prosecution. There is no substitution for a hands-on portion of any class. Chemistry and emergency medical technician (EMT) courses always have a laboratory element, because not everything can be learned by reading a book. You can learn how to apply a traction splint to a broken femur in EMT class, but until you actually apply such a splint, you only know what you are supposed to do, not how to do it. In the same way, the field portion of a fire investigation class is every bit as important as the didactic portion.

Along with fire investigation courses, it is valuable to take criminal justice courses. Almost every fire investigation you are ever involved in will eventually have some law enforcement aspect to it. The more you know, the better off you are. The courts have long recognized that a fire investigator can trample a citizen's constitutional rights as easily as a police officer can. Most firefighters don't see the need to be well versed in the legal aspects of the job, and sometimes that attitude extends to fire investigation. The fact that you do not carry a firearm or that you hold the title of "firefighter" rather than "police officer" will not save you from blowing up a felony arson case if you do not understand the law.

Many four-year universities now offer bachelor's degrees in fire administration, including the management of fire investigation bureaus. Although these programs are relatively new, they are excellent ways of gaining more education and therefore more credibility on the witness stand. It is important to take advantage of the good training and experience available to you.

Fig. 1–6. Community colleges are excellent places to begin your career in fire investigation. Many of these institutions offer specialized training and satisfy classroom requirements for prospective fire investigators.

Summary

To be a successful fire or arson investigator, you must invest a good deal of yourself in education and experience to gain the necessary credibility to assist in the criminal prosecution of arson cases. Fire investigators are often called on to act as both factual witnesses and expert witnesses and can only do this successfully through a combination of experience and training. Much of the time you will be expected to achieve both of these items on your own with little or no help from the fire departments that employ you.

Fire investigation is not pursued in the same manner by all fire departments, so as a fire or arson investigator, you must be well aware of your status within the department and also of the legal status of this position. While NFPA 1033 is a valuable and useful document, it does not specifically spell out what type of training or experience is enough to qualify as an expert witness in a court of law. That knowledge must be gained through interaction with the district attorney's office and the courts.

Several agencies and fire departments provide formal educational opportunities to new investigators and to those wishing to become fire investigators. This training will not come looking for the student, but the student must make every effort to seek out this formal training and education.

Experience may be more difficult to come by than formal classes. If you are a suppression firefighter, you may have the opportunity to watch experienced firefighters investigate fires and actually have these investigators teach you some of the tricks of the trade. If this isn't a possibility, you might have to look for a fire department that would be willing to allow you to work as an intern with the fire investigation unit. While this will require a good deal of work and time commitment from you, it will pay off in experience and training that you might not have been able to obtain otherwise. To obtain such a position, seek out the chief of the department and the commanding officer of the unit, and explain your needs and desires and how an internship might benefit both parties.

Continuing education is the cornerstone of becoming an investigator and remaining a successful one. As court decisions are rendered in our nation, the complexion of the job changes, and it is incumbent on you to remain current in your field. The only way to do this is through attending various seminars and college-level classes. The training will never cease, and your desire to learn should never wane.

Key Terms

Arson investigator. Generally, a member of a fire department or law enforcement agency who is responsible for investigating fires in which a criminal act is suspected. These investigators usually have police powers and the ability to make arrests. In some fire departments, arson investigators also act as fire investigators.

District attorney. An official responsible for representing the government; also called *the people* in court cases when prosecuting criminals. There is generally only one district attorney per jurisdiction, but several deputy or assistant district attorneys may also work under the DA's supervision.

Expert witness. A witness who by virtue of training, education, skill, or experience is deemed to have knowledge beyond that of the average person. This knowledge and experience are generally considered to be such that others may rely on expert witnesses' testimony, and their opinions may be considered in regards to evidence.

Factual witness. An individual who knows specific facts about the case in which he or she is testifying. These facts are generally things that the factual witness saw, heard, or otherwise witnessed. Factual witnesses are generally not allowed to venture any opinions during the trial.

Fire investigator. A member of a fire department, law enforcement agency, or even a private individual who is responsible for investigating fires to determine their cause and origin. When criminal acts are suspected, these investigators usually assist or concede the investigation to an arson investigator who is vested with law enforcement authority.

Fire investigation task force. In some smaller jurisdictions and during large incidents in large jurisdictions, fire and arson investigators from different agencies will form a group with the purpose of training for and investigating fires. Some fire investigation task forces enjoy a permanent status, and only the members vary.

International Association of Arson Investigators (IAAI). The IAAI is a professional association of individuals who conduct fire investigations. The IAAI provides resources for training, research, and technology. Membership is open to fire, police, insurance, and private investigators.

Interrogation. A form of questioning with a goal of uncovering information concerning a crime. It is generally considered to be more aggressive questioning than an interview, but the differences are really academic. The circumstances surrounding the interrogation or the interview may invoke constitutional issues.

National Fire Academy (NFA). One of two schools in the United States operated by the Federal Emergency Management Agency (FEMA) at the National Emergency Training Center in Maryland. It is operated by the United States Fire Administration and is considered to be the country's leading federal fire training and educational institution.

Ride-along. An arrangement made between an individual and an agency such as a fire department or police department to accompany firefighters or officers on their job for a specific period of time, to observe what they do and how they do it. Ride-along programs are usually formal programs that must be applied for and approved in advance.

Stamp. A slang term used in the fire service to describe a formal certification or document establishing that a level of training has been achieved.

Stamp collector. A slang term used in the fire service to describe an individual who spends a significant amount of time trying to collect as many certifications as possible to list on a resume.

V pattern. A mark on a wall or other flat object left by fire as it burns. These patterns are used by fire investigators to help in determining the area of origin of a fire.

Review Questions

1. Why is it important to gain professional certifications and formal education when attempting to become a fire or arson investigator?

2. What is the difference between a fire investigator and an arson investigator?

3. What are some of the advantages of seeking and receiving an internship with a fire department investigation unit?

4. What is the role of the district attorney in prosecuting a criminal case?

5. What is the major concern with NFPA 1033 regarding fire investigators' training?

6. Name at least two methods of gaining experience as a fire investigator before actually becoming a member of an investigation unit.

Activities

1. Visit a fire department in a community near you and find out how involved it is in fire investigation and what you might do with this department to further your education and experience in pursuit of a career in fire investigation.

2. Spend some time in a library or on the Internet to find what educational opportunities are open to you in the field of fire investigation. Visit the NFA's website and read about their fire investigation courses and other related classes.

2

Nothing Is Complete until the Paperwork Is Done

Learning Objectives

Upon completion of this chapter, you should be able to:

- Explain the importance of writing a clear, concise report.
- Describe some ways to improve report-writing skills.
- Describe the proper method of identifying witnesses, suspects, and victims for your report.
- Explain the importance of taking good notes in the field.
- Describe the roles sketches and photographs might play in a fire report.

Case Study

In 2006 I was living in Salem, Oregon, and working for a fire department as its deputy fire marshal. I was in the middle of an inspection when my cell phone rang. I immediately recognized the phone number as belonging to one of my buddies who still worked in the arson unit of the Long Beach Fire Department in California. As I answered the call, I hoped that he was just checking on me since we had not talked for a while, but I was pretty sure that wasn't the reason for the call.

I answered the call and was informed that I would need to return to Long Beach because a suspect I had arrested and convicted in 1995 had been arrested again, this time for murder by arson. My testimony was to be used under California law to establish past criminal acts that were related to the current charges. When I arrested this suspect in 1995, he had nailed shut the door to a motel room containing his girlfriend and then set the room on fire. I recalled the major parts of the case, including the result of his trial, but beyond that, 11 years had caused my memory of the details to fade somewhat. From past experience I knew that details were going to be important in this case, as the prosecution would be looking to establish a pattern of criminal activity for the suspect.

I told my friend that I would be glad to make the time to return to California and help prosecute, and ideally, convict this man, and that I was saddened to hear that a human being had lost his life due to this man's actions. Because the woman in the motel room had survived, he was convicted of a lesser crime, and therefore he served a lesser sentence. I also told him that I was very fuzzy concerning the details of the fire and arrest in 1995, and that I would need my reports to study before I testified. He said he had anticipated that, and the reports had already been faxed to my office. I also found out that I would be needed in Long Beach within a few weeks' time.

As it turned out, I ended up making two trips to Long Beach over the next year, and indeed I was able to testify twice. My testimony was partly responsible for putting this man back in prison for the rest of his life. I was able to give valid and truthful testimony in this case because I could study my well-written and accurate reports from the first investigation. Thanks to these reports, I was able to bring 1995 back into focus, after 11 years, and be helpful to the prosecution in this very important case. Without accurate and complete reports, this would not have been possible. Nothing is ever complete until the paperwork is done.

Introduction

You can be the smartest, best trained, and most experienced fire and arson investigator in the free world, but if you can't put your findings down on paper in the form of a report, you have a problem. Most district attorneys (and, for that matter, most chief investigators) want your investigation report done in a timely manner; they want your spelling, grammar; and syntax to make sense; and they want it done in a specific format that can be presented in court. They want you to include sketches of the fire scene, complete a photo log, and attach any and all lab reports—and they want it *now*. Most jurisdictions require that all suspects be charged within a very strict timetable or be released. That means that you must complete all of your reports within a short period of time for them to complete the **criminal filing** and to bring the suspect into court for his or her **arraignment**. The most common time frame is usually 48 to 72 hours from the time of the arrest. If you make the arrest on Friday night, Saturday, or Sunday, you will have a little bit more time, but you are still looking at having all of your reports due by noon or so on Tuesday. It is not unusual to take two or three suspects into custody over the course of a weekend, so you can imagine the workload this might put on you, the arresting officer. Even if you are not an arson investigator, even if you just work the fire investigation end of it, the person filing the criminal case will need your report and the same requirements apply.

You can also count on your report becoming a focal point for the defense. In most cases, your report is going to be the only professional account of the events that led up to the arrest of the suspect. Therefore, the defense is going to go over it with a fine tooth comb, scouring your report not only for inconsistencies in your description of the fire scene, but also for spelling errors, grammatical mistakes, and anything that could lower your status in the eyes of the jury. The argument will be that if you are not competent to put together a simple written report in a professional manner, how could you possibly be competent as an arson or fire investigator? It is a difficult argument to ignore or dispute.

It Is Never Too Late to Learn

It is never too late to become a better technical writer. Many community college fire protection programs require at least one, and in many cases several, writing classes as part of their degree programs. This is because many people are not well trained in writing technical reports. Fire and arson investigation reports are very technical reports, and to be an effective investigator, you must acquire these skills. The goal is to write a report so that 12 of your peers can understand what you saw, what was said to you, and, of course, your conclusions and how you arrived at them.

A report is just a story, and anyone can learn how to write a good story. The trick is to develop a procedure for creating a quality report each time. This chapter provides some guidelines and techniques that will help you write better reports. However, the need for formal education in writing cannot be emphasized too strongly. Writing detailed and technical reports is not instinctual. It takes time, effort, and formal instruction to learn to write well, and you must invest some time in learning to do so; otherwise, you are setting yourself up for disappointment.

It does not matter if you are the fire investigator who is going to hand the report off to law enforcement or if you are the arson investigator who is going to own this case all the way to the jury's final verdict. Your report needs to be clear, concise, understandable, and done in a professional manner. Keep in mind while you are writing your report that the goal is not to see how quickly you get the report done; it is rather to see how well you can do it within the given time frame you have to complete this report. This may sound daunting if you don't consider yourself a good writer, but there is a solution: Become a good writer.

Begin Your Report Early

In a sense, you start your report as soon as you arrive at the scene of the investigation. You should start gathering and storing information that you will later use back at the office or the fire station when you actually sit down to create the formal report. If you don't have the information you need when you write up the report, you will be in trouble. Think of it in classroom terms: It would be very difficult to write a term paper for any college class without first doing some research on your subject before you begin writing. Your notes, interviews, and observations at the fire scene are the research you need to complete your report (fig. 2–1).

Take careful notes at the scene. You cannot rely on your memory when you are gathering information that will eventually become a legal document. What might seem like a simple thing to remember while you're in the field might become the detail you need to pull your entire report together.

Fig. 2–1. Begin your report early. Take good notes and do complete interviews while you have the opportunity. A good field interview will yield a good investigation report.

Know Whom You Are Interviewing

You need to know exactly whom you are dealing with at a fire scene. Everyone you interview has the potential to be a witness, victim, or suspect if the case goes to trial. As an investigator, you are going to spend time talking to people who actually have something to say about the incident. If they are important enough to interview, they are certainly important enough to identify properly. The proper way to do this is not to ask them who they are, but ask for their identification. As the investigator at the scene, you have the legal right and the obligation to do this, and 95% of the people you talk to are fine with showing you their identification. The other 5% have something to hide, starting with their identity. Enlist the help of a uniformed police officer if you must, but secure their valid driver's license or government-issued identification card.

Let us take a side trip for a moment and discuss what you should accept as identification and what you should not. Realize that while not everyone has a valid state driver's license or identification card, about 97% of the people living in the United States do. It is very difficult to get a job, rent a house or apartment, rent a car, or do just about anything else without one. It would raise a red flag for me if someone told me they had absolutely no form of government-issued identification, and I would be very cautious about accepting other forms of identification, although there are a few that would work for me. A valid passport is one of those pieces of identification, and I would probably accept a valid military identification card. What I would not accept is any form of identification that does not have a photograph on it. Without a photo, you really have no idea whom you are dealing with.

Make Full Use of an Identification Card

Once you have their identification, use it to its fullest. Record their full true name including middle name, date of birth, identification number on the card, complete address, height, weight, and hair and eye color. You should also take great care to make note of their ethnicity, visible scars or tattoos, moles, birthmarks, and any other notable features that will help identify them in the future. When you are done, do not hand the identification back to them until you have had a chance to ask them all the questions you need to ask. Very few people are willing to walk away and leave their driver's license or identification card with you. It is also a good idea to ask for a second piece of identification with their name and picture on it if possible. There is a good reason for doing this: In some states it is easy to get a driver's license or identification under a false name. While it is not a common thing that people do, it is probably a lot more prevalent than you might think. Some people will keep a set of legitimate identification with them but use a false license for use in situations with law enforcement. Keep your eyes open for other names on cards in their wallets when they retrieve their identification for you, and see if they can back up the primary identification with another piece with the same name and picture on it.

The average person carries two to four pieces of picture identification, so it will not be hard for most people to comply. You can be a bit more lenient with the second piece of identification. What you might not accept as a primary piece of identification will work as a secondary piece of identification. Examine their identification with a little bit of skepticism. If the picture doesn't look right to you, then question them about the identification. Maybe even conduct a little verbal test such as asking for their middle name, date of birth, or their height. Sometimes you will discover that they will not be able to answer the question immediately, indicating that they might be using "borrowed" identification. If someone has to think more than a second or two about their date of birth, they may very well be using someone else's identification (fig 2–2).

Fig. 2–2. Know exactly whom you are talking to. Victims, witnesses, and suspects should all be carefully identified and their identity documented before you conduct an interview.

Once you are satisfied that you have received a valid identification card, put the card in your pocket or on your clipboard, but retain control of the card until you are finished asking questions. If they have a car with them, write down the license plate number including the state it is registered in, the make and model of the car, the year, the color, any visible body damage, and other identifying features. Ask for the vehicle registration papers and see if they match the name and address on their driver's license. If the name and address on the registration does not match up with the name and address on the driver's license, run the car registration number through the department of motor vehicles and find out exactly whom it belongs to. Then you can ask why the person you are talking to has it. If you don't have the ability to run vehicles and persons of interest for **wants and warrants,** have a police officer do it for you. There are several reasons for doing this, but the most important is it will enhance the possibility of finding a suspect or a witness at a later date. If you don't know whom you are looking for, or what kind of a vehicle they might be driving, they can be very difficult to find.

On the other hand, the proper identification of a suspect or witness can yield a wealth of information at a later date when researching information on your suspects and witnesses. Knowing a suspect's physical description, date of birth, driver's license number, and last known address can help to make sure you are dealing with whom you think you are. Having all of this information will allow you to ask the department of motor vehicles to provide you with more current information regarding a suspect's current location or a registered address for the vehicles they own. You can also have the department of motor vehicles provide you with a copy of the license it issued; in some cases, you will discover a completely different picture than that on the license you saw. While securing this information may require a search warrant, the information it yields will be well worth the time you spent gathering complete data on the suspect.

Organize and Standardize Your Note-Taking Procedures

When you are conducting interviews, learn to take good notes and to put them down in a concise, legible manner while you are still in the field. Keep two things in mind: These notes are the basis of the investigation and report, and these notes are discoverable by the defense attorney if the case goes to court. There are probably many ways to take good notes in the field and keep them organized, but many experienced investigators find one method to be particularly useful. They keep a small box in their vehicle containing a large supply of legal tablets, pens, pencils, a stapler, and some file folders. When they conduct any type of interview, they take a legal tablet from the box and begin by properly identifying any suspects, witnesses, or victims. They record all of the information they have secured (described at the beginning of this chapter) and then begin the interview. When they are finished taking these notes, they immediately tear out all of the pages concerning the interview, staple them together, and then put them in a file folder dedicated to this particular investigation. It is good practice to never begin a new interview until this is done, and never record information from a second interview on the same page as the first. Paper is cheap, pens are plentiful, and you are going to be held responsible for the manner in which you conducted and organized your investigation in the field. It is important to remember that all of your notes are subject to **discovery,** so you should write and organize them with this in mind (fig. 2–3). The prosecutor is going to read your notes along with your report; a defense attorney is going to read your notes; and the jury is more than likely going to know exactly what you wrote in your notes. Keep all of this in mind when you are writing them. Be thorough, be complete, and above all else be professional. Leave personal observations about suspects, witnesses, and victims out of your notes unless they have some bearing on the case. If your suspect is sweating profusely during the interview and seems much more nervous than he or she should, record it. Avoid at all costs making any personal comments about the person you are interviewing if it is not pertinent to the case. Any negative comment about a suspect might be brought up later during the trial phase, with the claim that you had prejudged his or her guilt based on an obvious dislike of the suspect. Do not allow that to happen.

Fig. 2–3. Organization in the field is the key to being able to write a clear, concise report. Spend some time to organizing your notes while you're still in the field.

Sketches of the Fire Scene

Sketches are quite useful because they provide an orientation of objects at a fire scene that photographs cannot (fig. 2–4). You do not have to be a graphic artist to create a useful sketch, and it doesn't have to be done by computer-assisted design. Sketches do not have to be drawn to scale, and they do not need to include every detail in the room, building, or vehicle. The entire purpose of the sketch is to provide some details about where you found the Molotov cocktail remnants, can of gasoline, or book of matches. A photograph of these items is, of course, critical, but plays an entirely different role in your report. A simple drawing might be able to speak volumes to a juror, and once again, it will demonstrate your professionalism and thorough investigation methods to the prosecutor, defense attorney, and jurors.

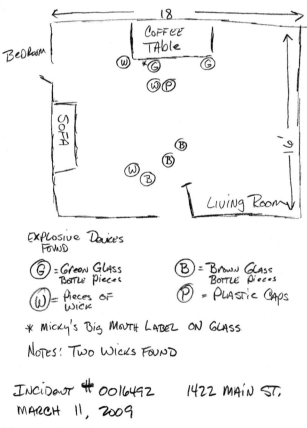

Fig. 2–4. Sketches are quite useful because they provide an orientation of objects at a fire scene that photographs cannot.

Photographs and Their Place in Court

Photographs taken at a crime scene are only admissible as exhibits when they are introduced by the person who took them along with verification under oath as to the authenticity of the photograph. You cannot hand a prosecutor a stack of prints and expect to have them admitted during a trial without your help. This may not be an issue if you investigate one or two fires a year, but when you are doing a dozen or so each month, one burned-out building or car begins to look like the next. You should take many photographs and use a photo log in the field to document them. Most prosecutors won't introduce 40 or 50 pictures at a trial, but will pick out five or six very important ones. The problem here is that you can't tell which ones will be selected while you are taking them. Keep in mind that you may not end up in trial on a fire until several years after the fact, and a matchbook lying next to a bed might not have as much meaning to you in 2013 as it did in 2006. There are many different ways to keep a photo log; find a method that works well for you and use it consistently.

A fire scene is not the proper place to learn to use your camera. Most departments encourage their investigators to learn how to use their work cameras in a variety of situations well before they are needed at a fire scene. Every time you use that camera, you will learn something about it and become more familiar with the equipment and its capabilities. Use your duty camera as much as possible and practice using it under adverse conditions. Think of it as training; the subject matter is really not important. It doesn't matter whether you are taking photographs of your nephew's birthday party, a baseball game, or a city park at night. It's better to find out that taking pictures in a very dimly lit room requires a particular lens setting when you are at a birthday party than at the scene of an arson.

Photographs, like any other pieces of physical evidence, will only be allowed to be introduced as evidence if the judge in the trial says they are admitted into evidence, and that only occurs when the photographer is present to testify under oath, and with the penalty of perjury, that the photograph in question is the one he or she took and that the photograph has in no way been altered.

Most fire departments and police departments now use digital photography for several reasons. The photographer can view the image immediately rather than waiting to see if the picture came out well after film processing. The expense of digital photography is so much less than film and processing that it becomes an important economic issue to most departments. Finally, it is becoming more difficult to buy and maintain good film cameras. Digital photography is here, and it is here to stay.

This brings up the issue of how easily digital photography can be altered and how easy it would be to fake a photograph at a crime scene using readily available photography programs on a computer. Keep in mind that the human element is still required in court, and no photograph is going to be admitted without the photographer being there to testify as to its authenticity. Digital photographs can be altered, as can film-based photographs, but to introduce any photograph into evidence in court, the photographer must raise his or her right hand and swear under oath that the photograph is genuine. Very few people are willing to commit a felony and attempt to introduce an altered photograph of any kind in a court of law.

So who could take a photograph and have it entered into evidence during a criminal trial? Would you have to be a certified fire or arson investigator with specialized training in photography? The simple answer is no, anyone could potentially have a photograph entered into evidence during a criminal trial. If a neighbor was to photograph an arsonist running from a burning house while taking pictures of her son's birthday party, that photograph could easily be entered into evidence under the same conditions described above.

Recording Video for Courtroom Use

All of the rules that apply to still photography also apply to video photography. Even the most basic cellular phones have the ability to take both still pictures and video. As an investigator, it is incumbent upon you to consider if such video or photos might have been taken by bystanders and might prove valuable to your investigation. Once again, it is important that you fully identify who took these pictures and video. Also, keep in mind that a video of a fire scene does not under any circumstances take the place of still photography. The reasons for creating a video are quite different. A video can be used to show the jury how the investigative process worked at the fire scene. Maybe you have an accelerant-sniffing dog, and you want to show the jury how the dog alerted you and your investigative team to the presence of a flammable liquid where it should not be. This type of video can have a dramatic effect on the jurors, and of course almost everyone loves dogs. This may sound like a little bit of showmanship, but that really is a major element in the courtroom. As in life, the side that does the best job of gathering all the information together and presenting it in a clear, understandable manner usually carries the day.

Time Is of the Essence

Now that you have gathered all of your observation notes, interviews, sketches, photo logs, and the run report, it is time to actually create the final report. Time is of the essence. There are a couple of reasons that time is a major factor. The first one should be obvious; the longer you wait to write a report about anything, the less you remember and small details are lost. You may take great notes, but the more time that passes, the less those notes will jog your memory. I believe I was questioned nearly every time I testified as to when I wrote my report. It's always better to create the formal report as close to the time of the investigation as possible (fig. 2–5).

Fig. 2–5. Once you have gathered your field notes, lab reports, and photos, it is time to create a clear, concise report.

The second major factor involves the Constitution and the filing of criminal charges regarding a suspect you have arrested or caused to be arrested. The majority of jurisdictions allow you to hold someone for about 48 to 72 hours before formal charges must be filed and the suspect arraigned. Whether you are the arresting arson investigator or the fire investigator working with the police department, this time frame is critical. If charges have not been filed at the end of this time, the person in custody will be released. As you can imagine, it becomes very difficult to find a suspect the second time. Being arrested for a felony tends to put suspects on edge, and they have a tendency to make themselves scarce. Time is critical, but a poor report can never be the result of time constraints. You must learn to do your job well, but to do it quickly.

Writing the Report

There are probably dozens of different formats that can be used for filing a clear, concise investigation report. Most fire and police departments use a reporting format that is somewhat unique to their department. You should always adhere to the approved format for your agency and always make your report as complete as possible. This was covered earlier in this chapter, but it cannot be stressed enough. If a little information is good, a lot of information is great. Many departments now use computer programs with pull-down menus for boiler plate information concerning the fire you are investigating. To some degree this is good as it saves time when you are entering information such as the owner of the property, the time of the fire, or the number of units responding. It is a great idea to enter that type of information using these menus. Unfortunately, some agencies have a tendency to try to capture too much information in this manner. Many departments will try to kill two birds with one stone, and they will be tempted to use the information captured during a fire investigation for other purposes. This is not a serious problem if your report is not going to be used to file a criminal case against an arsonist, but a fire investigation report that is going to be used to charge someone with arson should not be used to capture statistical data. Questions that are used to compile numbers such as "first material ignited" or "area of origin" should not be part of an investigation report when limited choices are provided on a pull-down menu. It's great to capture this information, but it should be done on a separate form.

A good fire investigation report should remind you of a short term paper written for a college class. There should be an introduction and the scene should be set. There should be a narrative including interviews with victims, witnesses, suspects, and subjects. Finally, there should be a conclusion including your opinions and observations.

The introduction is pretty straightforward and should go something like this:

> At 1400 hours on May 2, 2008, I, Investigator Dave Johnson of the Centerville Fire Investigations Unit, was dispatched to 2874 Riverdale Way in Centerville to investigate a house fire. After my arrival, I was briefed by Captain Sam Smythe concerning the nature of the fire and actions taken by the suppression crews.

This should be short and sweet, about a paragraph long. It gives the readers all the information they need about who you are, where you were, when you were there, and what you were there to do.

Writing the narrative is pretty much a matter of style, but there are some things you can do to make your report more professional and less of a challenge for the prosecutor. It should go without saying that proper sentence structure, syntax, grammar, and spelling are a must. Have at least one other person proofread your report at least twice, and try to select a proofreader who has writing skills at least equal to yours, and preferably better. You cannot do an effective job of proofreading your own report. You are too close to the subject, and as you read your report, your mind knows what you are trying to say even if it might not be clear to others. This is a critical step in report writing but is often overlooked. Fire departments are known for employing people from all walks of life as firefighters; somewhere in your department there is probably a firefighter who went to college to become a schoolteacher and then changed his or her mind. Make use of that person's area of expertise and ask him or her to proofread your reports (fig. 2–6).

Fig. 2–6. Find someone to proofread your reports for you before you submit them to anyone inside or outside of the department.

Computers Are Useful, but Not Perfect

All word processing programs have a version of **spell check**, and most of them will find basic spelling and grammatical errors. That should cut down on the time and make your report easier to correct, but even this useful program has its limitations. The problem is that sometimes you will use a word in the wrong place, but the computer will not pick that up because the incorrect word you used is indeed a word and might be properly used in a different sentence. The computer cannot determine what you meant to say; it can only tell whether you the words you used were spelled properly. Here is an example: Let us assume that during a fire investigation you found a gasoline can on the *dock* near a fishing vessel that burned. But when you wrote your report you made a typographical error and stated you found the gasoline can on the *deck* near a fishing vessel that burned. This will not be picked up by the word-processing program because dock and deck are both valid words and spelled properly. It will, however, change the location where you found the evidence and could later be difficult to explain in court. Even if you can explain it away in court, it will put doubt in the mind of some jurors about your ability to accurately write a report. So take advantage of technology and use it to its fullest, but before you sign and turn that document over to another agency, make sure it is correct and accurate.

Stay Within Your Writing Abilities

It is important that you do not try to write above your abilities. Use words and phrases that you would normally use in your everyday speech. Keep it real and don't use a word like *perfidious* when you would normally use a word like *untruthful*. Likewise, if you do have an extensive vocabulary and you normally use it, write as you speak. There is no reason to "dumb down" a report. You are who you are, and the jury and judge will be able to tell if you are dumbing down a report. That can be either good or bad for you, but you won't find out until it is too late. You should always be yourself, write as yourself, and assume that everything will come out fine. This is not to say that you should not continually strive to improve your writing skills; you should. As you do, you will write better reports because you have indeed become a better writer. Nothing is more obvious and nothing makes investigators look worse to the jury than using words in their reports that they really don't understand. Once again, be yourself in your reports, but become a better writer.

It is perfectly fine to refer to yourself as "I" or "me" in your report. In the first paragraph of your report you identified yourself, so the reader knows who you are. Another suggestion is to keep everyone you mention in your report at an arm's length. In other words, if you have a victim named Reginald Kenyon, do not refer to him as Reg or Reggie or even Reginald in your report. You should refer to him as "the victim" or simply as Mr. Kenyon. This goes for everyone mentioned in the report. You should not refer to fire department officers as Jim or Sarah; refer to them as Captain Salisbury or Chief Bingham. Reports are formal, legal documents, and you should always remember that while you are creating them. There is an old saying that familiarity breeds contempt, and the last thing you want or need is the jury regarding you in that manner.

Your Job Is to Draw Conclusions

Finish your report with a conclusion and your opinions. Many arson investigators don't think they should venture an opinion as to the cause of a fire, but that is exactly what we are paid to do. Your opinion should be based on facts gathered by a thorough and scientific investigation, but in the end, most conclusions you come to will simply be your opinion. If you have concluded that the fire was an act of arson set by human hands using a match or other open-flame device, then say so. Being a fire or arson investigator can be a very confrontational job. You will be placed in situations where your investigations, reports, and opinions are going to be used to put people in prison for a very long time. Presumably, you have been granted the position of fire or arson investigator by your department because of your training and abilities. It is your job to decide who did what and when. You must always remember that you know more about the fire investigation than anyone else in the courtroom. Once you are installed as your department's investigator or you are established as an expert, your opinion is specifically why you are there.

Honesty Is Always the Best Policy

This is a good time to talk to you about **exculpatory** statements and evidence that you might discover that could conceivably assist the defense, should the case come to trial. It is absolutely critical that you adopt the philosophy of taking the good with the bad. You follow the evidence wherever it might lead. If the evidence you discover proves your suspect to be innocent, then so be it. There are a couple of reasons for living by this rule. The first and most important is that no arson investigator or police officer wants to send an innocent person to jail. As an investigator, you need to remain impartial as to the outcome of an investigation, and you need to seek the truth above all else. This is a basic tenet in the emergency services field. We are there to help people, and for no other reason. By discovering the truth about a fire investigation and making sure that innocent people are not accused of crimes they did not commit, you are helping everyone.

As you might expect, you are going to get a lot of conflicting information during a criminal investigation. Some of this information is going to be circumstantial or conflicting because fire scenes are by nature difficult crime scenes to work. A lot of evidence is going to be destroyed or moved around by fire streams and firefighters doing their jobs. It doesn't take much for 1¾-inch hose to blow pieces of a beer bottle used as a Molotov to the wrong side of the room. Our job is not to return these items to where we think they might have been. Our job is to locate, recognize, and record the evidence accurately, and then explain in a scientific manner why the items are where they are. Our job is to remain completely objective during our investigation and follow the evidence wherever it leads us. This applies even when the evidence leads us somewhere we did not expect and that we might not be particularly happy about.

Personal Agendas Do Exist

Some of the conflicting evidence or statements are going to be outright lies by people who have something at stake in the case. This doesn't just include the suspect, but people who will purport themselves to be witnesses. Parents, other relatives, and boyfriends or girlfriends will often provide you with outright lies in hopes of helping their friend or relative. Sometimes they may believe they have something at stake even when they don't and will not be truthful with you. An example is a fire in which you believe a child was playing with matches. The child may have even told you he or she was playing with matches, but the parents might dispute this for fear that their homeowner's fire insurance won't cover the loss. Record their statements accurately and include all other evidence and statements you have gathered disputing their claims. Remember that you are not there to help them file a claim with their insurance company, and you are not there to judge them. You are there to write a clear, concise, and accurate report concerning your investigation and what you saw, smelled, and heard. Omitting information that is important to your case for fear of harming someone is as bad as putting false information into the report for the purpose of harming someone. Your job is to do a complete and scientific investigation and then to pass an accurate report along to the governing agency.

People Do Make Mistakes

You are also going to interview people who really believe they saw or heard something that just wasn't possible in your estimation. That does not mean they are lying to you, and it does not mean they are trying to hide something from you. Fires are chaotic, and the average person is not exposed to them on a daily basis as you are. They become excited, and in that excitement, they sometimes honestly believe they have heard or seen something that just isn't possible. Listen to them very carefully, and take what they say into consideration, but if what they said they saw or heard is not possible, it must be noted in your report. Anyone who has ever attended a baseball, football, or basketball game knows that perfectly honest people disagree on what they saw or heard. You may disagree with an umpire concerning a ball or strike call on any single pitch, but it doesn't mean one of you is a liar. Sometimes it is just a matter of perspective, and this occurs at fires as well.

The point is that you do not have the option of selecting evidence and only including items in your report that would support a conviction of the suspect. It is incumbent upon you to write your reports honestly and without an agenda other than the truth. Ideally, you will investigate every statement anyone makes to the best of your ability; either you will strengthen your case by proving that untruthful witnesses are lying or you will discover the truth and start looking in another direction for the actual arsonist. Either way, it is a win–win situation for you. The last thing you want to do is to put a good case in jeopardy because you failed to include interviews in your report or you failed to follow up on a lead because it didn't suit your goals. The average defense attorney is going to find these people and use your lack of professionalism against you even if the witnesses couldn't have hurt your case. Be thorough in your investigation and just as thorough in your reports.

After You Have Completed Your Report

Once you have completed your fire investigation report, print a hard copy, sign it, and then file the report along with photos, lab reports, sketches, and any other pertinent documents. Ideally, you have a secure file room where there is a reasonable expectation of the file being there when you need it. Saving a copy of the report in **PDF** format on your computer is fine, but it is no substitution for a signed hard copy. Computers crash or get stolen, files get deleted accidentally (or on purpose), and a computer savvy person can open and change a PDF format to an alterable document. Defense attorneys know this, and you can count on it coming up at some point if you use a PDF rather than a signed hard copy.

At this point, your report is now a legal document and subject to discovery during a hearing or trial. The same goes for your field notes, photographs, sketches, and any other documents you have created. More importantly and really the whole reason for writing the report, it is now the instrument through which you can educate the prosecuting attorney and refresh your memory for testimony at a later date. As our case study showed at the beginning of this chapter, you will never really be sure that you are finished with a case and that you will never need to review the report again. Fire and arson investigation is a field in which vast amounts of time can pass with no action, and then out of the blue a case can come back to haunt you. The best way to prepare for this possibility is to write good reports that will be as useful in 10 years as they were 10 days after the incident.

Summary

Report writing is one of the most important aspects of being a fire or arson investigator, and the value of a good report cannot be overstated. There is no secret to writing a good report; all it takes is hard work and a desire to produce a good report for future reference.

If your writing skills are not what they should be, take a writing class at a local community college. Even if you were once a good writer but are out of practice, this type of class can prove valuable in achieving your goal of producing a report that will stand up in a court of law or with your fire chief.

Make sure you take advantage of all that technology has to offer such as word-processing programs that will check your spelling and grammar, but remember that the human touch cannot be replaced. Have someone else proofread your report and look for errors that might not have been picked up by a computer program. What you wrote may make perfect sense to you, but will others understand what you are saying?

Develop a method for gathering information at a fire scene, and make sure that you identify everyone you come into contact with at the fire scene. This will prove valuable when attempting to contact witnesses or find suspects at a later date. Make sure you carry a large supply of legal tablets, pens, pencils, sketch books, and other equipment necessary to take notes and make other observations at the scene of the fire. When you think of writing your report, think of starting it as soon as you get to the fire scene. As good as you believe your memory is, there is no substitute for good notes to assist you in writing the report.

Organize your notes while you are at the fire scene, and then write your report as soon as possible. Time can be a critical factor for a number of reasons. Most jurisdictions allow only a 48- to 72-hour window to file reports and to charge a suspect with a crime related to an arrest. It is also important to write reports in a timely manner before the events of the investigation begin to fade from memory.

Learn how to take photographs at a fire scene and how to sketch a fire scene. It is not necessary to become a professional photographer, but it is important to learn how to use your equipment in advance. In addition, sketching fire scenes is an underused skill and one that can paint an overall picture for the jury.

Be objective in your report writing, and begin your investigation without an agenda. Follow the facts where they lead, and then write your report in the same manner. Make sure that you include all information in your report in a fair and honest manner.

After writing your report, you need to print, sign, and store it in a manner that will preserve your work. It may be years before you need to refer to your report, and during that time, computers will crash or be replaced. There must be some type of hard copy archives of all of your investigations.

Key Terms

Arraignment. The first appearance of the accused before a judge in a criminal proceedings, wherein the charges against the accused are read to him or her and his or her plea is recorded. During arraignment, a lawyer is appointed if the defendant cannot afford one.

Criminal filing. Also called a criminal complaint, this filing formally charges the person named or an unknown person with a particular offense.

Discovery. Part of the pretrial litigation process, during which each party requests relevant information and documents from the other side in an attempt to "discover" pertinent facts. Generally, discovery devices include depositions, interrogatories, requests for admissions, document production requests, and requests for inspection.

District attorney. An elected official of a county or a designated district with the responsibility for prosecuting crimes. The duties include managing the prosecutor's office, investigating alleged crimes in cooperation with law enforcement, and filing criminal charges or bringing evidence before a grand jury that may lead to an indictment for a crime.

Exculpatory. Applied to evidence that may justify or excuse an accused defendant's actions, and that will tend to show the defendant is not guilty or has no criminal intent.

PDF. Portable document format (PDF) is a file format that provides an electronic image of text or text and graphics that looks like a printed document and can be viewed, printed, and electronically transmitted.

Spell check. An electronic dictionary in a word processor that can be used to catch misspelled words.

Wants and warrants. A term used by law enforcement referring to what jurisdiction might "want" a suspect, or what suspect might have an arrest warrant outstanding. An arrest warrant is a document that calls for someone to be brought into police custody. Arrest warrants are official orders from the state calling for the accused to be held to answer for a specific crime. If an arrest warrant is issued against someone, they will be taken into custody as soon as the police locate them.

Review Questions

1. What are the most important reasons for writing your fire report in a timely manner?
2. What would be a good way to improve your writing skills?
3. At what point in the fire investigation should you consider beginning to gather information for your report?
4. What form of identification is considered to be the most acceptable when identifying a suspect or witness?
5. What type of information would you want to include in your report with regard to identifying the person being interviewed during a fire investigation?
6. What is normally required when submitting a photograph to the court with a motion it be entered into evidence?
7. What useful role does a sketch of the fire scene provide when you are filing your report?
8. What is one of the dangers of relying only on computer programs to check the mechanical accuracy of your reports?
9. What is the normal time window allowed for an investigator to file a report when it is the basis for an arrest?

Discussion Questions

1. Explain the many reasons that it is important for the fire or arson investigator to produce a well-written and accurate report.
2. Explain why it is necessary to have a second or third person proofread reports before they are submitted to the district attorney or fire chief.
3. Explain why it is important for a fire investigator to write about all the information and evidence of the investigation and avoid cherry-picking facts in the report.

Activities

1. Visit a local community college with a copy of a report you have written. Ask an English teacher to read your report and make some suggestions about how you could become a better report writer.

2. Visit your local fire station, and ask if there are some fire reports you could read. Make sure the reports are not being used in any active fire investigation cases, and then look at them and see if you feel that they were written in a clear, concise manner. Think about what changes you would make to improve the reports.

3

Case Law and NFPA 921

Learning Objectives

Upon completion of this chapter, you should be able to:
- Describe the importance of the Fourth Amendment of the U.S. Constitution and its impact on fire investigations.
- Describe the impact of *Michigan v. Tyler* on the investigator's right to conduct a fire investigation.
- Describe the circumstances under which suspects must be advised of their Miranda rights.
- Describe what might constitute custodial interrogation and what would not.

Case Study

It was 0300 hours when Investigator Bob Compton arrived at the scene of a house fire in an upscale neighborhood. As was his habit, Compton began his investigation by obtaining a briefing from the battalion chief concerning the fire. The chief told Compton that arson was suspected in the fire because the flames flared back up after being doused with a hoseline, and the neighbors had seen a man running from the structure after the fire had started. Compton asked the chief if there was a description of the man, and the chief replied that firefighters had actually found the man and he was now sitting on the back of the fire engine, guarded by a couple of firefighters. Compton thanked the chief and then walked to the back of the engine, where he found a man, appearing to be about 35 years old, sitting on the tailboard and two firefighters standing with him.

Investigator Compton approached the suspect and introduced himself, stated that he was there to investigate the fire, and asked him if he knew anything about the fire. The suspect stated that the house belonged to his girlfriend and that they had broken up in the last week. He stated that he had become engaged in an argument with her earlier in the evening and had left. The suspect further stated that he returned shortly after midnight and found her there with another man. He stated that he had become angry and had driven to a gas station and bought some gasoline, using a metal can he had in the back of his car. Compton asked him what occurred after that, and he stated that he had broken a window in the rear of the house, poured the gasoline in the window, and then threw a match in after the gas, setting the house on fire.

At this point Investigator Compton asked the suspect to accompany him into the house and show him where the window was and identify the metal can he had used to buy the gasoline. The suspect complied, and after taking the metal can as evidence, Compton informed the suspect that he was under arrest for arson and took him to jail to be booked.

Compton wrote his fire and arrest report that night, and the next morning, he took them to the district attorney to file criminal charges against the suspect. After reading his reports, the district attorney looked up at Compton and said, "This is very interesting, but I don't see in your report where you read the suspect his rights. Did you forget to include that?"

A sick look came over Investigator Compton's face, and he replied, "No, I never read him his rights because the fire chief detained him rather than me." The district attorney looked at Bob and said, "Sorry buddy, but none of this is admissible. Go let your suspect go; I am not filing charges."

Knowing the Law Is Our Responsibility

Many firefighters and most fire investigators talk about *Michigan v. Tyler*, but most investigators know very little about the case. The same goes for *Miranda v. Arizona* (1966), *Mapp v. Ohio*, and *Rhode*

Island v. Innis. As a fire or arson investigator, you must become familiar with these important Supreme Court cases because there is a good chance that nearly every investigation you ever conduct will be affected by one or more of these cases. You must also become familiar with and understand the Fourth and Fourteenth Amendments to the United States Constitution. You will deal with these constitutional amendments that almost daily as an investigator. To be true professionals, we must have a firm grasp on their impact on every part of our professional lives (fig. 3–1). The example cited in the case study above is not particularly far-fetched and, as a matter of fact, is based on a real instance. Failure to understand statute and case law can have a negative impact on our ability to do our jobs well, and those failures can affect people in a personal way that may carry through the rest of their lives. In the case study, the victim of the fire must now live with the knowledge her boyfriend is probably not going to be prosecuted for setting her house on fire, and she will live in fear of him as he is still walking the streets.

Fig. 3–1. Every fire investigator should read and understand the U.S. Constitution. To do otherwise could negatively affect your career.

The days are gone when we as firefighters or as emergency medical technicians (EMTs) can say, "Well, that is a police matter and nothing for me to concern myself with." The old saying that what you don't know can't hurt you could not be further from the truth. What you don't know could cost you a very important arson case and could easily have a negative impact on your career.

The Fourth Amendment

It is critical that all firefighters, fire investigators, EMTs, and law enforcement officers fully understand their rights and obligations in relation to investigating the fire scene. To do otherwise is to invite liability on yourself and your fire department. The owners or occupants of a fire-damaged property have a right to privacy and due process of law. These privacies are protected by specific amendments in the United States Constitution

The Fourth Amendment states:

The right of the people to be secure in their persons, houses, papers, and effects, against unreasonable searches and seizures, shall not be violated, and no warrants shall issue, but upon probable cause, supported by oath or affirmation, and particularly describing the place to be searched, and the person or things to be seized.

The Fourth Amendment was born of the English belief that every man's house is his castle. The most famous case involving search and seizure is *Entick v. Carrington*. In this case, state officers with general warrants raided many homes and seized pamphlets critical of not just the king's policies, but the king himself.

The issue in colonial America at that time centered not so much on seditious libel issues but seizures of goods deemed "prohibited and unaccustomed" by the Crown. To aid them in collection of revenue, the English issued writs of assistance, which were general warrants authorizing the bearer to enter any house or other place to search for these items. Once issued, the writs remained in force for the lifetime of the sovereign plus six months. This meant that a representative of the British government, most likely with soldiers, had the right to enter anyone's home at any time of the day or night and without any particular reason. They didn't need to be looking for anything in particular, and there was no such concept as **probable cause.** A knock, or a boot, would come at the door, and the occupants were obligated to allow the soldiers into their home as if the soldiers actually had a reason to be there. If the soldiers decided they were hungry or tired, then they had the right to eat the occupants' provisions or to occupy the home as if it were an inn. The soon-to-be Americans considered this invasive practice such an important issue that the Third Amendment to the Constitution dealt with the quartering of soldiers. As one might conclude, these searches did not sit well with the Americans, and that gave rise to the Fourth Amendment to the Constitution which is part of our **Bill of Rights.**

The Fourteenth Amendment

Section One of the Fourteenth Amendment states:

All persons born or naturalized in the United States and subject to the jurisdiction thereof, are citizens of the United States and of the State wherein they reside. No State shall make or enforce and law which shall abridge the privileges or immunities of citizens of the United States; nor shall any State deprive any person of life, liberty, or property, without due process of law; nor deny to any person within its jurisdiction the equal protection of the laws.

There are four other sections of the Fourteenth Amendment dealing with the apportionment of representatives, serving in the house or senate, public debt, and Congressional authority to enforce the provisions of the article. By far, the first section is the most important to the fire investigator. The purpose of the Fourteenth Amendment is to guarantee equal treatment to all citizens of the United States regardless of race, creed, ethnicity, social standing, or wealth. The most famous cases of all to arise for the Fourteenth Amendment are *Miranda v. Arizona*, which produced the ubiquitous Miranda warnings, and *Mapp v. Ohio*. These cases will be discussed later in this chapter.

Michigan v. Tyler

Everyone likes to talk about *Michigan v. Tyler*, but few actually understand what the court's decision means to firefighters and to fire investigators. Failing to properly understand this critical Supreme Court decision can jeopardize a fire investigation in two specific ways.

First, you run the risk of violating someone's constitutional rights. You must understand that even the person you believe to be an arsonist enjoys constitutional rights, and you must be extremely careful to respect those rights and do nothing that might place those rights or your case in danger. Evidence that is seized improperly might prevent an arsonist from being prosecuted and punished for his or her crime and instead going free and a victim of that crime receiving no justice. It doesn't matter how guilty you believe the person is, it doesn't matter how much you dislike them, and it doesn't even matter if they are citizens of the United States or here legally. Constitutional rights belong to everyone in the United States. Second, if you do not understand *Michigan v. Tyler*, you could easily place restrictions on your activities and unnecessarily burden your investigation by bestowing rights on the **perpetrators** they do not merit by the Constitution or by the Supreme Court decision. Understanding the case will help to keep you from doing either of these things, because neither is desirable.

Shortly before midnight on January 21, 1970, a fire broke out in a furniture store belonging to Loren Tyler and Robert Tomkins. The Oakland County (Michigan) Fire Department responded, and as they were extinguishing the flames, the fire chief arrived, around 0200 hours. Following extinguishment, discovery of containers of flammable liquid was reported to the chief. He entered the building to examine the containers and summoned a police detective to investigate the possible crime of arson. The detective took several pictures but ceased further investigation because of heat and smoke. By 0400 overhaul was complete and the firefighters departed. The chief and the detective removed the containers and left the fire scene to allow the smoke, heat, and steam to dissipate.

At 0800 the chief and assistant chief returned, made another examination of the scene and removed further evidence. On February 16, 1970 (about three weeks after the fire), a state police arson detective conducted an investigation of the fire. The investigation included photographs and removal of evidence. Tyler and Tomkins were charged with arson and conspiracy, and they were subsequently convicted in a court of law.

The Michigan Supreme Court overturned their convictions based on the Fourth and Fourteenth Amendments. On appeal by the state, the U.S. Supreme Court then reversed the decision of the Michigan Supreme Court and reinstated the convictions.

In its decision, the U.S. Supreme Court stated the following on May 31, 1978:

> We hold that entry to fight a fire requires no warrant, and that once in the building, officials may remain there for a reasonable period of time to investigate the cause of the blaze. Evidence of arson discovered in the course of such investigations is admissible at trial, but if the investigating officials find probable cause to believe that arson has occurred and require further access to gather evidence for a possible prosecution, they may obtain a warrant only upon a tradition showing of probable cause applicable to searches for evidence of a crime. We do not believe that a warrant was necessary for the early morning re-entries on January 22. As the fire was being extinguished, Chief See and his assistants began their investigation, but visibility was severely hindered by darkness, steam, and smoke. Thus they had departed at 4:00 am and returned shortly after daylight to continue their investigation. Little purpose would have been served by their remaining in the building, except to remove any doubt about the legality of the warrantless search and seizure later that same morning. Under these circumstances, we find that the morning entries were no more than an actual continuation of the first, and the lack of a warrant thus did not invalidate the resulting seizure of evidence.

The court also found that the entry of the state police arson detective on February 16 was clearly detached from the initial exigency and constituted a warrantless entry and search of the premises. All evidence obtained after the January 22, 1970, investigation of Chief See, his assistant, and the police detective was excluded.

The state's case was substantially buttressed by the testimony of former employee Oscar Frisch. Mr. Frisch testified he helped Tyler and Tomkins move valuable items from the store and old furniture into the store a few days before the fire. He also related that Tyler that told him there would be a fire January 21 and ordered him to place mattresses on top of other objects so they would burn better.

There are several important concepts that we can take away from this court decision that will aid us in doing better, more complete investigations. Understanding this case will also help us to avoid making our investigations more difficult by requiring unnecessary steps before conducting an investigation. Everyone agrees that it would be ridiculous to require firefighters to call a judge and get a warrant before being allowed to enter a structure and extinguish a hostile fire that was destroying the structure. Very few if any legal scholars would mount an argument stating that the fire department is violating the Fourth Amendment rights of the homeowner or tenant by making a warrantless entry into a building to extinguish a fire. The existence of **exigent circumstances** is a well-established legal precedent and is applicable not just in the case of a burning building but in several other situations as well. The Supreme Court in *Michigan v. Tyler* recognized this but also went a step further. It said that once inside the building, officials had the right to conduct an investigation into the cause of the fire without benefit of a warrant, and any evidence they found could be seized and would be admissible in a court of law.

There are a couple more points concerning *Michigan v. Tyler* that need to be discussed, given that many fire departments and fire investigators have a distorted view about what *Michigan v. Tyler* says about continuing the investigation and maintaining control of the building in question. There is a misconception that as long as the fire department leaves an engine company on the scene of a fire or hires a security guard to stand near the building, *Michigan v. Tyler* remains in force and they control the scene for as long as they desire; this is simply not true. What the decision actually said was that officials may remain on the scene for "a reasonable period of time" to conduct an investigation into the cause of the fire. What has never been definitively established is what constitutes a reasonable period of time. In some jurisdictions fire investigators are on the scene of a fire within 15 minutes of the fire being extinguished, while other jurisdictions make take several days to launch an investigation. Once on the scene, some investigations can be conducted in a matter of a couple hours, while more complex investigations may take several days or weeks to come to a conclusion. What is reasonable will undoubtedly be determined by another Supreme Court case someday in the future.

Another **caveat** concerning *Michigan v. Tyler* is the part dealing with gathering evidence of arson and its admissibility in a court of law. It would clearly appear that the court was stating that any evidence found in **plain sight** and in the immediate vicinity of the fire would be admissible in a court of law. This does not mean that *Michigan v. Tyler* gives fire investigators the right to conduct searches of the property without a warrant, nor does it allow investigators to seize evidence that does not meet the requirements stated above. In fact, *Michigan v. Tyler* clearly states that if the investigator finds probable cause that a crime has been committed and a further search is necessary, the investigator must follow the proper procedure for securing a search warrant before proceeding with the search.

Michigan v. Tyler is a very useful tool for fire department investigators and one that should be completely understood and utilized. This is as necessary for fire investigators to be able to conduct an initial investigation into the cause of the fire as it is to protect the rights of the victims or even the suspects.

Clifford v. Michigan

A home belonging to Raymond and Emma Jean Clifford of Detroit, Michigan, was damaged by a fire at around 0540 hours on the morning of October 18, 1980, while they were out of town on a camping trip. Firefighters were able to knock the fire down quickly and were finished with extinguishment shortly after 0700 hours. After extinguishment, all fire and police personnel secured the operation and left the Cliffords' home. They also left what they thought might be a timing device in the driveway of the Cliffords' home for the arson investigators when they finally reached the scene.

Approximately five hours later, at about 1300 hours, Detroit arson investigators arrived at the residence for the first time to investigate the cause of the blaze. They found a work crew on the scene boarding up the house and pumping about 6 inches water out of the basement. Investigators were informed that the Cliffords had been notified of the fire and had instructed their insurance agent to send the crew to secure the house. Without benefit of a search warrant, the investigators entered the residence and conducted an extensive search without the warrant and without getting the permission of the Cliffords. Detroit arson investigators found a crockpot cooking device with two bottles of propane fuel attached to it and what they believed to be a timing device. It was determined this device was the cause of the fire and it was seized in the basement and was marked as evidence. Investigators continued to search the Cliffords' home without obtaining a warrant, and their search led to the upper floors of the house, where additional evidence of arson was discovered. The Cliffords were arrested and charged with arson, but moved to suppress all the evidence seized in the warrantless search on the grounds that it was obtained in violation of their rights under the Fourth and Fourteenth Amendments. The Michigan trial court denied the motion, stating exigent circumstances justified the search. On appeal, the Michigan Court of Appeals found that no exigent circumstances existed, and it reversed the lower court's decision.

In the example of *Clifford v. Michigan*, it should be obvious that the situation was far different from the situation presented in *Michigan v. Tyler*. In *Clifford*, the fire department had left the scene of the fire and had actually turned the residence over to a representative of the Cliffords. At that point the status of the property changed, the exigent circumstances were plainly lacking, and the Cliffords had every right to expect to be protected by the Fourth Amendment of the Constitution. The fire investigators should have been aware of this, and they should have acted accordingly. No exigent circumstances existed as they did in *Tyler*, and the investigators did not have permission from the Cliffords or from their representative to enter the property, conduct an investigation, or conduct a search of the house in areas not involved in the fire. Absent permission to conduct the investigation and the subsequent searches, the investigators had no right to enter the property. In addition, any permission the investigators might have secured would have to have been granted by the Cliffords themselves or their representative, the insurance agent. Obtaining permission from the board-up crew to conduct an investigation or a search would not have been sufficient.

In this case the investigators probably had the right to stop the work being done on the house if they felt it was destroying evidence of a crime. At that point they should have contacted the Cliffords and their representative and attempted to secure permission to conduct a fire investigation and a search of the property. It is very likely that permission would have been denied, and if so, the investigators should have then contacted a judge and requested a search warrant using whatever probable cause they had.

In fairness to the investigators in Detroit, this was only about two years after the *Michigan v. Tyler*. However, as an investigator, you should make a concerted effort to stay abreast of changes in the law.

To do otherwise might lead to the unhappy circumstances described above.

Mapp v. Ohio

Mapp v. Ohio once again concerns a violation of the Fourth and Fourteenth Amendments to the Constitution. This time the appellant, Miss Dollree Mapp, prevailed with the aid of the American Civil Liberties Union. On May 23, 1957, police officers burst into the Cleveland, Ohio, home of Mapp and her 15-year-old daughter. As an excuse for entering the home of Mapp, the Cleveland police stated they had "information" that a person wanted in connection with a bombing was hiding in the house. After calling her attorney, Mapp demanded to see a search warrant. A police officer held up a piece of paper, and Mapp grabbed the piece of paper and shoved it into her bra. A struggle ensued, and the "warrant" was recovered. Mapp was arrested for belligerent behavior, and the search revealed no bomb paraphernalia, but a lot of books, photographs, and pictures considered to be lewd and lascivious. Mapp was convicted of the belligerent behavior charges but appealed her conviction based on a violation of the Fourth and Fourteenth Amendments.

The U.S. Supreme Court overturned Mapp's conviction, and all evidence gathered in the illegal search was excluded. The State of Ohio contended that even if the search was conducted without authority or otherwise unreasonable, the state was not prevented from using the unconstitutionally seized evidence at trial. They further went on to claim that the Fourteenth Amendment does not forbid the admission of evidence obtained by unreasonable search and seizure.

The court disagreed and stated that the exclusionary rule is an essential part of both the Fourth and Fourteenth Amendments. They further said that the state, by admitting evidence unlawfully seized, serves to encourage disobedience to the federal Constitution which is bound to uphold. The criminal goes free, if he or she must, but it is the law that sets him or her free. Nothing can destroy a government more quickly than its failure to observe its own laws, or worse, its disregard of the charter of its own existence.

Mapp v. Ohio was the originator of the concept of **fruits of a poisonous tree,** and it provides an important lesson for all investigators regardless of their discipline. Even though this case had nothing to do with an arson or fire investigation, it is important for fire investigators to realize that any evidence seized under less than legal conditions will not be allowed to be introduced into evidence at trial. There are legally established methods for fire and arson investigators to follow to conduct investigations and to seize evidence of arson when they find it. It is our responsibility to become familiar with these methods and follow them to the letter.

If you go back to the case study at the beginning of this chapter, it should occur to you that anything Investigator Compton learned from his suspect would never be admitted into a court of law. Even evidence that he may have recovered later might not be admitted if the defense could show that the investigator would never have discovered it without help from the suspect.

Miranda v. Arizona

No case is more widely known or more widely misunderstood than that of *Miranda v. Arizona*. In 1963 Ernesto Miranda, an eighth-grade dropout with a criminal record, had been picked up by Phoenix, Arizona, police officers and accused of raping and kidnapping a retarded 18-year-old woman. After two hours of police interrogation, Miranda was never told he had the right to remain silent, to have a lawyer, or to be protected from self-incrimination.

The confession was presented at Miranda's trial; he was convicted and sentenced to 20 years in prison. His appeal went all the way to the U.S. Supreme Court, and the rest is history. This was the birth of the **Miranda rights** that police officers, fire arson investigators, and others in positions of authority are required to issue to arrestees before questioning.

The ruling offered only a temporary reprieve to Ernesto Miranda. He was retried without the confession and convicted once again based on evidence and testimony of a former girlfriend. He served 11 years before being paroled in 1972.

In 1976 Ernesto Miranda was fatally stabbed in a bar fight. The police arrested a suspect who, after being read his Miranda rights, chose to remain silent. The suspect was released, and no one was ever charged with the killing.

For the record, the most widely accepted version of the Miranda warning is as follows:

> You have the right to remain silent. Anything you say can and will be used against you in a court of law.
>
> You have the right to an attorney and to have an attorney present during questioning. If you can't afford to hire an attorney, one will be provided at government expense.
>
> At this point it is wise to secure an affirmative reply to two questions and to note the questions and answers in your report.
>
> Do you understand the rights I have just read to you?
>
> Having these rights in mind, do you wish to waive these rights and speak to me?

If both of these questions are answered in the affirmative, you are on strong legal grounds to question your suspect, and the odds are in your favor that any statement made at this point will indeed be admitted during trial (fig. 3–2a).

There is no law that requires the Miranda warning to be in written form or recorded by any electronic device. Written and signed warnings are a matter of departmental policy. I do not believe that requiring a written Miranda warning signed by the suspect is a particularly good policy (fig. 3–2b). If we accept for the sake of argument that it is the police officer's or arson investigator's responsibility to educate the criminal as to his or her constitutional rights, I would ask what is the purpose of putting the suspect on an even sharper edge right before questioning? I have heard the argument that now we can "prove" we issued their Miranda rights. Never once in my career was my word under oath that I had Mirandized a suspect questioned. Telling someone something is one thing, but having them sign a paper reinforces what Miranda implies: never cooperate with law enforcement, and above all else, never incriminate yourself. I personally do not believe it should be a police officer or arson investigator's job to help a criminal avoid prosecution. I don't think anyone should be deprived of their rights, but if you are going to be a decent criminal, you should educate yourself.

It is my duty to warn you before you make any statement that:

1. You have a right to remain absolutely silent.
2. Anything you do say can and will be used against you in a Court of law.
3. You have a right to consult an attorney before making any statement.
4. If you are without funds you have a right to a Court appointed attorney at public expense.
5. You have a right to have your attorney present when and if you do make any statement.
6. You have the right to interrupt the conversation at any time.
7. Anything you do say must be freely and voluntarily said.

(over)

Fig. 3–2a. Every fire and arson investigator must be familiar with the Miranda rights and their origin. This is the standard Miranda warning.

DO YOU UNDERSTAND THESE RIGHTS?

DO YOU HAVE ANY QUESTIONS ABOUT YOUR RIGHTS?

Date: _____

Signature acknowledging constitutional rights:

_____

Officer's Signature: _____

Fig. 3–2b. On the back of most Miranda cards is a place for the suspect to sign. Doing this is a matter of departmental policy.

Experienced officers will keep a copy of the warning on a small card in their pocket and read the warning each time rather than repeat it from memory. This point is often brought up at trial, and investigators are asked if they read the rights to the suspect of if they recited them from memory. The defense attorney is looking for an opportunity to claim that the rights might not have been properly administered if they were recited rather than read, and therefore the defendant was not afforded the opportunity to hear a proper warning. Experienced investigators will put their initials and identification number on the

Miranda warning card and have it in their pocket at trial. Do not be surprised if the defense attorney asks you to produce the card and read it aloud in court. What the defense is hoping for is that you will not have the card in your possession, and then he or she can make the point that maybe you didn't have it with you when you made the arrest. Here is how you can avoid that; obtain many Miranda cards and put copies in all of your suit jackets or uniform coats or dress shirts. Make a practice of reading the card rather than reciting it. That way there will be no mistakes, and you can tell the absolute truth in court and look good doing it. It is often little things like this that will instill confidence in the investigator for the jury, while the lack of these little things that will plant a doubt concerning your competence.

Miranda is widely misunderstood, and television has caused us to believe many things concerning the *Miranda* decision that are just not true. The fact is that *Miranda* is only important and it only applies when two things take place:

1. A suspect is placed in custody.
2. The suspect is questioned or interrogated.

The problems usually arise when investigators or other fire personnel who are not investigators do not understand these two concepts and violate the suspect's rights without realizing it. While these two concepts seem straightforward, there are many nuances to *Miranda* that we need to understand.

The definition of custody in relationship to *Miranda* will nearly always come up at trial (fig. 3–3). The questioning of a suspect outside *Miranda* can be a lucrative area for the defense to start, as any mistake in this area may very well lead to the rejection of all other evidence, statements, or confessions that arose from the interrogation phase of the investigation.

It is very important to remember that *Miranda v. Arizona* does not just apply to police officers. The actual wording in the decision is "persons in positions of authority." It has also been interpreted to include those acting as agents of persons in positions of authority. An example of using an agent would be an investigator sending a relative of a suspect into a locked room at police headquarters, so that the relative could question the suspect on behalf of the investigators; absent a Miranda warning, it is likely none of those statements would ever be admissible in court (fig. 3–4).

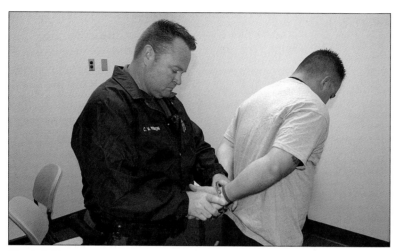

Fig. 3–3. To even the casual observer, it is obvious that this situation would create a custodial situation for the arson investigator.

Fig. 3–4. After carefully reading this suspect his rights, the investigator had him sign a Miranda card to remove all doubt as to whether he had received his rights or not.

Placing someone under arrest is certainly a form of custody, but there are other custodial situations that do not rise to the level of arrest. If a fire chief were to detain an arson suspect or even someone the chief thought was an arson witness in his or her vehicle until the investigator arrived, it would likely be a custodial situation, and *Miranda* would apply. If a police officer or fire investigator were to question the same person, there is a distinct probability that the resulting statements would be inadmissible in a court of law (fig. 3–5).

Fig. 3–5. It is virtually guaranteed that the rights of the suspect are going to be challenged later in a courtroom setting. The investigator must be ready to testify about his or her actions.

Conversely, if an investigator were to question an individual on a street corner during his or her investigation and the individual were to make incriminating statements during that interview, these statements would be perfectly legal and admissible in a court of law, as no custodial situation existed.

So the giant question is and will always be, what constitutes having someone in custody? Unfortunately, there is no standard and consistent answer to that question. Clearly, if you put someone in handcuffs and deprive him of his freedom by placing him in the back of a patrol car, he is in custody. But what if you "ask" a person to sit on the tailboard of a fire engine and wait to talk to the investigator? What if you ask one of your larger firefighters to keep an eye on the person? What if you question someone in her living room and you are between her and the door? The answer is that all of this is decided on a case-by-case basis, and each judge will make his or her own determination about what constitutes custody (fig. 3–6).

The most important thing to take from this discussion is an awareness of what *Miranda* is really about. The goal is not to avoid a possible custodial situation, but to recognize one so the arriving investigators and police officers can be advised of the situation. Many good arson cases have been lost because of a lack of communication among the participating parties. It is much better to create a custodial situation and deal with it than to allow an arsonist to disappear forever.

Fig. 3–6. The situation here is less clear. Could the suspect be judged to be in custody in this situation? It is certainly something the fire investigator should consider.

On the subject of questioning, do not let *Miranda* intimidate you while gathering information during your investigation. Simple questions concerning a witness or potential suspect's name, address, date of birth, and the like, are not subject to *Miranda* to begin with. Noncustodial questions relating to the fire investigation are perfectly legal, and in fact, it is incumbent upon a good investigator to ask them.

Rhode Island v. Innis

Sometimes in an attempt to outsmart our suspects we end up outsmarting ourselves by stepping over a line into something called **functional equivalent.** Stepping over that line can easily ruin a perfectly good arrest and conviction by violating *Miranda*, in spirit if not the letter of the decision.

Thomas J. Innis robbed and murdered a taxi driver in Rhode Island. He was arrested after a second robbery, during which he had used a shotgun. The shotgun was not recovered by the arresting officers. Innis was given a *Miranda* warning, and at that time he invoked his right to have counsel present during questioning. On the way to the police station, the officers who were transporting him talked among themselves about the need to find the shotgun because they were near the location of a school for handicapped children. The officers during their conversation said they were concerned that one of the children might find the gun and hurt themselves or another child. Innis, overhearing the conversation, volunteered to show the officers where he had hidden the shotgun.

At his trial, Innis's attorney sought to suppress his statements in which he revealed the location of the shotgun, on the grounds he had been subjected to custodial police interrogation after invoking his right to have an attorney present. The U.S. Supreme Court noted that the trial judge believed the officers were being truthful when they testified that they had not intended to provoke Innis into talking when they discussed among themselves the need to find his gun. The court also held that the officers could not have anticipated that Innis would volunteer to show them where the gun was hidden. The court ruled that this conversation between the officers in the car did not amount to an interrogation.

Innis is important because as investigators we must understand that interrogation is not limited to asking suspects direct questions while they are in custody. Basically, functional equivalent means that you cannot do or say anything once a suspect is in custody that would cause the suspect to make incriminating statements unless that suspect has been advised of her Miranda rights. For example, assume you are transporting a suspect to jail to be booked after he has been arrested. Since you have no intention of interrogating the suspect while you are driving, you have not advised him of his Miranda rights yet. During the drive, he asks you if you would like to know what happened, and you nod your head up and down, indicating that you would like to hear what he says. In all likelihood, this is going to be seen as the functional equivalent of questioning outside *Miranda* and will probably be inadmissible.

The proper response to the suspect's question would be to pull the car over to the side of the road and then tell him you would love to hear his side of the story, but first you must advise him of his rights. Advise him of those rights, and then let him talk. Other forms of functional equivalency might be sending a relative into a jail cell to question the suspect about a fire while you listen or carrying on a conversation with your partner in the presence of the suspect with the intention of tricking her into making an incriminating statement.

Voluntary Statements

Nothing in *Miranda* forbids officers from hearing and noting what are known as voluntary statements from suspects. Suspects do sometimes blurt out voluntary statements that can be incriminating in a court of law. On television you see actors portraying police officers throwing their hands over their ears and shouting for the suspect to shut up because they cannot hear these voluntary statements. In real life, this does not happen. For example, you could be transporting a prisoner from the booking area to the jail in an elevator with your partner when the prisoner blurts out a confession such as, "All I did was burn her car!" Let us assume that this statement was made after the suspect was Mirandized and after he refused to speak to us or answer any questions. This statement would still be admissible in court. Even if his attorney attempted to have the statement excluded because his client had refused to answer any questions without an attorney present, it would very likely be unsuccessful as no attempt had been made to question the suspect, and the statement was a **voluntary utterance**. It is possible that the suspect in this example might have watched too much television and was under the impression that nothing he said could ever be used in a court of law. In this particular example, he would be wrong.

Noninterrogational Questions

It should be clear that nothing in the *Miranda* decision is intended to keep you from asking questions concerning the identity of your suspect. Standard booking questions, custodial instructions, and inquiries that are normally asked after an arrest are not considered to be interrogation, even if they happen to provoke an incriminating response. Only speech and actions that could predictably prompt an incriminating response are considered to be interrogation. The courts have stated that only words or actions that you know or should have known would likely elicit an incriminating response are considered to be interrogation. Here is a real-life example of how this could occur: After investigating a fire bombing, two members of the arson unit and I took four young men into custody. That is, they were arrested, handcuffed, and transported to jail. When I arrested the suspect I was with, I advised him of his Miranda rights, and he chose not to talk to me. When we arrived at the jail, I removed him from my car, and he said, "Can I ask you a question?" I responded by saying, "Yes, of course." He then looked at me and said, "Do you think what I did tonight will keep me from becoming a police officer?" This is commonly known as a confession and was used against him at trial. His attorney argued it was outside *Miranda*, but the court correctly ruled there was no questioning of the suspect and no functional equivalent. He could just as easily have asked me what time breakfast would be served in the morning. By the way, this did keep him from becoming a police officer.

NFPA 921

As a fire investigator, it is important for you to realize that National Fire Protection Association (NFPA) 921, *Guide for Fire and Explosion Investigations,* is *not* the standard for fire investigation; it is a *GUIDE*. The word "guide" is in the title rather than "standard." A technical standard is an established norm or requirement. It is usually a formal document that establishes uniform or technical criteria, methods, processes, and practices. A guide is something that offers basic information or instruction. It is very important that you under-stand the difference between a guide and a standard and keep it in mind at all times.

By contrast, the title of NFPA 13 is *Standard for the Installation of Sprinkler Systems.* NFPA 13 in an adopted standard in Chapter 9 of the International Fire Code. Because it is a standard and because it has been adopted by reference, it bears the full weight of law, and therefore it is the way we install sprinkler systems.

NFPA 921 is a good guide for all fire investigators to study in order to gain some idea about how to conduct a fire investigation. However, it should not be the only tool at your disposal for learning how to conduct what amounts to a criminal investigation. Ideally, you take several formal classes before venturing into the field, and if you are fortunate, you shadow an experienced investigator at several fires before conducting your own investigations under the watchful eye of a mentor.

NFPA 921 is a consensus document. That means it is only what a number of committee members could agree upon, but by no means the final word on the correct or incorrect way to conduct a fire investigation. There is generally more than one way to do anything, and in most cases, there are several good ways to accomplish any task. Fire investigation is no different, as each investigator has a little different way of doing things, and each way is correct in its own way. For example, when you arrive on a fire scene to conduct an investigation, you might want to touch base with the incident commander or the captain who was first on the fire scene. You might want to let him or her know where you are going and what you are going to be doing and find out from the incident commander what phase of the operation suppression was in. Also, if you can limit the amount of crime scene contamination caused by overhaul, it is a good thing.

Other investigators like to do a complete walk around of the exterior of the building and form a loose, preinvestigation opinion of what they expect to hear from the incident commander. Their approach might be that they want to view the physical aspects of the scene before hearing input from anyone else. Both methods have good and bad points, and I certainly wouldn't state under oath that one is right and one is wrong.

In some cases, NFPA 921 may pose some problems and challenges for fire and arson investigators. The NFPA is an outstanding organization and produces more than 300 useful and important standards and guides. NFPA 921 is one of those guides, but it is not the only book, document, or guide that you should use to learn how to conduct a fire investigation.

In most cases that eventually go to trial, and in almost all civil cases you might become involved in, you are probably going to be **deposed**. A deposition is a statement taken under oath in advance of a criminal or civil trial. It is an investigative tool used by attorneys. One of the common things that attorneys will look for at trial is a deviation in statements made by an investigator or an inconsistency in the way the investigation was conducted. Good defense attorneys will make certain that fire and arson investigators adhere to all fire department policies, all laws involving arrest and interrogation, and what is considered to be acceptable investigative practices.

During the deposition, it is common for attorneys to ask fire and arson investigators if they are familiar with NFPA 921 and if they adhere to the practices outlined in NFPA 921. Many defense attorneys believe that NFPA 921 provides them with the official document they need to catch you in what they will characterize as a mistake in your investigative procedure. A word of caution: The defense attorney may attempt to get you to concede or agree that NFPA 921 is the gold standard by which all fire investigators should live and that any deviation from NFPA 921 represents pure and unadulterated incompetence on the part of the fire investigator. You do not want to agree to that thesis under oath, in the hallway, or in any report that you have ever written. You don't want to do this for two reasons: First, it is not true. NFPA 921 is not a referenced or adopted part of the International Fire Code or any other set of laws I am aware of. Many investigators who do use NFPA 921 as reference material also use many other publications for the same purpose. Second, no one—and I mean no one—can adhere to any publication under all conditions without exception, 100% of the time. It just isn't possible. You must understand that the entire reason that the defense attorney is even bringing this up is to lay the groundwork to discredit the testimony of the fire or arson investigator at trial, by finding something in NFPA 921 that contradicts what you did at an investigation after you swore under oath you always followed it and it is the standard for fire investigators.

Understand that the investigator they intend to discredit with your statements is you. It is a safe bet that your departmental policies and procedures or standard operating procedures will at some point come into conflict with something in NFPA 921. It might be the way you take or store evidence, or it might be the manner in which you file reports or how you maintain those reports after you write them. But at some point you are going to have to choose between making a false statement under oath because you said you always follow NFPA 921 or admit under oath that you violated your department's procedures. Neither of these choices is a good thing that will enhance your case at trial.

So how do you avoid this? It is very easy. When you are asked about NFPA 921, you should answer truthfully. You should state that you have read NFPA 921, that you are familiar with the document, and that you consider it a good guide with some valuable information, among several other publications. State that it is not by any means the standard for conducting a fire investigation and that there are several things in it that you might vary from because of departmental policy. At this point, be ready to cite at least one or two important points in NFPA 921 that you disagree with or that are in conflict with another well-known publication. Don't set yourself up for a later embarrassment on the witness stand. Become familiar with what NFPA 921 is and isn't; someday you will be glad you did.

Once again, though NFPA 921 is a good and useful publication, it is not the only one, and it is by no means the standard when conducting a fire investigation. The NFPA itself does not refer to it as a **standard,** but correctly calls it a **guide.** It is important that you as a fire investigator understand the distinction and are clear in those distinctions once you are under oath.

Summary

As citizens of the United States we all have rights that are guaranteed by the Constitution. These rights are not just extended to law-abiding citizens, but to all citizens, foreign nationals in our country legally, and even those who are not in our country legally. These rights extend to everyone regardless of their status in our country.

The Fourth and Fourteenth Amendments to the Constitution are among the most important and most basic rights that we enjoy. They were written to ensure our right to be secure within our homes and from unreasonable search and seizure, and to make sure that everyone enjoys these rights.

Case law is made when specific cases come to the attention of our highest court in the land, the Supreme Court, and its decisions become law through precedent. There are several cases that have had great impact on the fire service and will continue to have great impact. It is our responsibility to understand these laws and to follow them to the letter. We are held to a very high standard as fire investigators, and we must strive to meet those high standards at all times.

Key Terms

Bill of Rights. The Bill of Rights are the first ten amendments to the U.S. Constitution and generally deal with individual rights.

Case law. The decisions and interpretations made by judges while deciding on the legal issues before them, which are considered as the common law or as an aid for interpretation of a law in subsequent cases with similar conditions.

Caveat. A warning or proviso of specific stipulations, conditions, or limitations.

Deposition. The testimony of a party or witness in a civil or criminal proceeding taken before trial, usually in an attorney's office. Deposition testimony is taken orally, with an attorney asking questions and the deponent (the individual being questioned) answering, while a court reporter or tape recorder (or sometimes both) records the testimony.

Exigent circumstances. An emergency, a pressing necessity, or a set of circumstances requiring immediate attention or swift action. Exigent circumstances allow law enforcement to enter a structure without a warrant.

Fruits of the poisonous tree. The principle that prohibits the use of secondary evidence in trial that was culled directly from primary evidence derived from an illegal search and seizure. Under the fruit of the poisonous tree doctrine, evidence is also excluded from trial if it was gained through evidence uncovered in an illegal arrest, unreasonable search, or coercive interrogation.

Functional equivalent. The principle by which one accomplishes something through an alternative act to accomplish the same ends. While an investigator might not technically ask a suspect a question, nodding one's head when asked if the investigator wishes to know something would be the functional equivalent resulting in the end.

Guide. Something, such as a pamphlet, that offers basic information or instructions advising a way of doing something.

Miranda rights. The rights that are read or recited by police in the United States to criminal suspects in police custody, or in a custodial situation, before they are interrogated.

Perpetrator. One who commits a crime or other offense.

Probable cause. The reasonable belief that a person has committed a crime.

Standard. A written document considered by an authority as a basis of comparison.

Voluntary utterance. A statement made without any questioning or prompting. It is a legal principle that allows statements made by suspects without questioning to be admitted during criminal trials.

Review Questions

1. What two circumstances are necessary before suspects need to be read their Miranda rights?
2. Who has the right to the protection of the Bill of Rights?
3. What does the Fourth Amendment guarantee?
4. What did the Supreme Court state a reasonable period of time to conduct an investigation is in *Michigan v. Tyler*?
5. Why don't firefighters need to get a warrant to suppress a hostile fire?
6. What is NFPA 921?
7. Why is it so important for fire and arson investigators to understand and abide by case law?

Discussion Questions

1. Even though it has nothing to do with fire investigation, discuss why the U.S. Supreme Court case of *Mapp v. Ohio* has an important impact on fire and arson investigators.
2. While NFPA 921 is a useful and important document, discuss how not understanding the difference between a guide and a standard might become an issue during a court trial.
3. Discuss what difference *Michigan v. Tyler* has made in the way fire departments conduct investigations.

Activities

1. Research literature by Devallis Rutledge and other noted authors on *Miranda* and other case law that have the potential to impact our profession. Discuss this literature with other investigators as it relates to investigators' work in the field.

2. Contact local fire investigators and find out what their departmental policies are concerning custodial interrogations of arson suspects.

4

The Participants in a Trial

Learning Objectives

Upon completion of this chapter, you should be able to:
- Explain the difference between a criminal trial and a civil trial.
- Explain the responsibilities of the district attorney and the defense attorney.
- Explain the role of public defenders.
- Describe what happens at an arraignment and at a preliminary hearing.
- Explain the role of the grand jury.

Case Study

In my life experience, I have never come into contact with anyone even remotely as powerful as a judge in a courtroom. Not my drill instructor in the Marine Corps, not the chief of the fire department, nor any principal or dean I ever faced in high school or college. When you step into a courtroom, the man or woman in the black robe becomes the most powerful person in your world. This power is virtually without limit, and the oversight of that power is very difficult to access and very constrained. The United States is a common law country, and as such, judges have powers that do not exist in other legal systems. One of these is the "contempt of court" power. A judge has the power to summarily imprison or fine anyone for any misconduct that occurs in the courtroom. Furthermore, the definition of misconduct varies from judge to judge and on a nearly daily basis. Contempt of court can range from insulting the court by using vulgar words or actions to merely showing up late to court.

I once witnessed a defense attorney who incurred a sizable fine for contempt of court because he showed up 45 minutes late for a court hearing the day before Thanksgiving. When asked for an explanation for his lack of punctuality, he explained that he was helping his girlfriend prepare a turkey for Thanksgiving dinner and was unable to leave on time. When the judge told him she did not feel that this was a particularly valid reason for keeping the court and all the participants waiting, he stated "Yeah, well you don't know my girlfriend." This was in my experience the most expensive turkey I had ever heard of. The defense attorney was also made aware of how fortunate he was to be able to enjoy it with his girlfriend the next day rather than having Thanksgiving dinner with the county. This incident occurred during the first six months in my position with the arson unit, and it has stuck with me for the rest of my career. Judges are not the right people to displease.

Introduction

A courtroom can be a terribly intimidating place if you are not familiar with the workings of the system or if you happen to be the defendant. Both of these things can be avoided by a little clean living and some preparation. But like any other workplace, once you understand what is going on in a courtroom, it becomes less intimidating, and with a little luck, you will eventually become close to comfortable in the setting. This chapter aims to familiarize you with some of the general parts of the court system and give you an overall understanding of the players and their parts. Once again, there is no substitute for experience, but understanding what you are supposed to be experiencing should help a great deal.

Criminal and Civil Trials

It is important to first understand that there are two basic types of court and types of trials: criminal and civil. If all goes well for you as an investigator, you will be much more involved in criminal court than in civil. It is, however, impossible to avoid either if you are going to be a fire or arson investigator. I would venture to say this goes for emergency medical technicians and police officers as well. When you do any of these jobs, you insert yourself into other people's lives, and that sometimes ends up with us involved in the court system. Since you cannot avoid it, the best thing to do is to understand it and prepare for it.

Criminal Trials

A **criminal court** deals with the prosecution of suspects who are accused of committing specific crimes. Generally, these criminal courts are divided into two categories: one that handles felony crimes and one handles misdemeanor crimes. In most large states, these two categories are known as superior court and municipal court, but the terminology varies from place to place. In some smaller communities, felony and misdemeanor trials might be handled by the same court. There is also federal court, which is tasked with dealing with crimes that have violated federal laws such as the deprivation of civil rights. It is possible that one of your arson cases could eventually lead to federal court.

Superior courts or state courts generally deal with **felony** crimes. Felony crimes are the most serious crimes and generally carry punishments from one year in prison to the death penalty. Arson is considered to be a felony crime in every state in the United States. There are lesser crimes involving fires that one could be charged with, such as reckless burning or failure to control a fire, that are misdemeanors or infractions, but arson is a serious crime, and in my experience, it is generally charged as a felony.

In some states there is a lower court that deals with **misdemeanor** crimes and **infractions**. You may very well become involved in a trial at the lower level. Many cases that start out as arson cases may end up being charged as misdemeanors for any number of reasons. The major difference is that in a misdemeanor trial, the stakes are somewhat lower as generally the upper limit of the punishment is one year or less in jail. Note here that there is a major difference in going to jail and going to prison. We will discuss that at a later time in this book.

Civil Trials

Civil court is the other type of court that we are going to discuss. There is an excellent chance that you will end up in civil court someday if you are an arson investigator and a pretty fair chance of going there if you are a fire investigator. Civil courts are reserved for private parties to attempt to legally find relief for damages they feel they have incurred at the hands of another party. The state or the people are not involved in civil trials from the standpoint of representing or defending one of the parties. The state is there only to provide a venue for the parties to have a judge or jury rule on their claims and differences. Large entities such as corporations or manufacturers may find themselves in civil court if they are named in a **tort.** To summarize and understand the major difference between a criminal and a civil trial: In a criminal trial a law has been broken, and the victim is represented by a district attorney or the equivalent. In a criminal trial the defendant can be fined, can go to jail, or can go to prison. In a civil trial two private parties disagree on some point and have decided to have a judge or jury settle their dispute. All that is at risk is money; jail or prison is not a possibility.

If you need a perfect example of this, you need look no further than the O.J. Simpson trials. Simpson was acquitted in his criminal murder trial, but held responsible for the deaths of his wife and Ron Goldman in a civil trial. This illustrates two things: (1) Civil court does not constitute **double jeopardy,** and (2) a good outcome in a criminal trial doesn't always guarantee a good outcome in a civil trial. The rules are different and so are the procedures. There is one other concept that you should understand about criminal and civil court. It is entirely possible that criminal charges could arise out of what is discussed or revealed in civil court. The reverse is also true and is probably the case more often. Evidence and testimony revealed during a criminal trial could easily result in a civil case being filed by one or more participants. What is said in court is, for the most part, a public record and can be used both for you and against you at a later proceeding. This is a very good thing for you to remember when you testify in court under oath.

Who's Who in the Courtroom?

Now that we understand the two types of trials, let's talk about how a criminal case comes to be such and who the players are. The first and most important link in the filing of an arson case or any criminal case is you—and I mean this literally. If you do not do your job, if you do not investigate the fire in a proper manner, and if you do not identify and interview suspects and witnesses, there will be no arrest and criminal filing. If you do not write your reports in a clear, concise, and timely manner, there will be no criminal filing. If you cannot explain the case to a deputy district attorney or a grand jury in a way they can understand, there will be no criminal filing. The entire case actually rests entirely on your shoulders; no one is going to do it for you. Now that we understand that the arson investigator is the cornerstone of any arson case, let's look at the different people and professions you will be working with.

The District Attorney

Your first stop will be the district attorney's office. The **district attorney** (DA) is a lawyer who is elected to represent the people in criminal cases (fig. 4–1). In larger jurisdictions there are also a number of **deputy district attorneys** (DDAs) or assistant district attorneys (ADAs). These assistants are not elected positions, but are hired to review arrest reports, to decide whether to file criminal charges against those who have been arrested, and to prosecute these cases after they have been filed. If you are fortunate, you might work in a jurisdiction that has a DDA who is well versed in fire and arson cases and understands the needs for the arson investigator. Otherwise, it is up to you to educate the DA so he or she can file the case. The federal equivalent of a DA is a U.S. attorney, each of whom is appointed by the president. They are assigned to regional offices and supervise assistant U.S. attorneys.

Fig. 4–1. Your first stop when filing a criminal case is the district attorney's office. A deputy district attorney will represent the people and prosecute the case on their behalf.

Your job in relation to the DA is to seek a criminal filing with regard to the person you just arrested. In other words, you have the suspect in custody based on probable cause, but now you need to bring formal charges against the suspect so that the case can proceed through the legal system. The rule of thumb is you have about 48 hours to put your investigation and arrest report together, along with supporting documents such as witness statements, and make your presentation to the DA's office. Filing a criminal case for the first time can be a bit intimidating and confusing, but once you learn the procedure for filing a case, you will become comfortable with it. After you have explained the entire case to the DDA, he or she will decide specifically what criminal charges should be filed and will write a criminal filing for you to take to the stenographer pool to be typed in the proper format for the court. After the filing is typed up, you need to sign it, and it is then taken to the court's office for filing. Please note that you will be signing the filing, not the DA The case is proceeding based on your investigation and report, so you are one actually filing the criminal charges. This is a great responsibility and could certainly place you in a precarious position if you allow it to. Right above the line where you attach your signature the following wording appears:

I DECLARE UNDER PENALTY OF PERJURY THAT THE FOREGOING IS TRUE AND CORRECT AND THAT THIS COMPLAINT, CASE NUMBER NA000000, CONSISTS OF THREE COUNT(S).

Then it has a line with your name typed underneath that identifies you as the declarant and complainant. Note that the warning above is printed in capital letters on the charging sheet, and perjury is a felony. I think what you can take away from this is the need to be absolutely truthful and precise on all of your reports. To do otherwise is to place yourself in jeopardy.

The Arraignment

After you have successfully filed criminal charges, the suspect will be arraigned. An **arraignment** is a court appearance in which the defendant is formally charged with a crime or crimes. The defendant is asked to respond by pleading guilty, not guilty, or **nolo contendere**. The arraignment is also when legal counsel is appointed for the defendant or the defendant tells the judge that he or she has an attorney. The trial date is usually set at the arraignment, and the question of bail is addressed (fig. 4–2).

Fig. 4–2. All parties will be represented by counsel, and any witnesses or interested observers are usually welcome to watch the proceedings.

Concerning the possible pleas at the arraignment, there are three possibilities. A **guilty plea** means just that: The defendant admits to the crime and stands ready for sentencing at the hands of the court. You shouldn't look for this to happen very often. Defendants are rarely allowed to plead guilty at their arraignment by their court-appointed attorney or the attorney they have hired. There really isn't any reason for a suspect to do this, and no defense attorney should allow it. About 98% of the time the suspect will plead **not guilty** at the arraignment and allow the process to move forward from that point. Once in a great while, someone will plead nolo contendere, but even this happening at an arraignment is extremely rare. If a defendant pleads nolo contendere, it means they are neither admitting to the crime nor denying that they committed the crime. They are, however, agreeing to accept the punishment for the crime as if they were found guilty. The advantage of this type of plea is it cannot be used as an admission of guilt later in a civil case if one were to be brought against the defendant.

Postarraignment

After the arraignment, the legal proceedings will proceed in one of two directions. A lot of what happens next depends on where you are in the country and how large the jurisdiction is in which you live. Politics and philosophy concerning prosecution of criminals will sometimes play a key role in the next step as well. After the arraignment, the case will proceed to either a **preliminary hearing** or a **grand jury**, both of which have the same objective, to determine if there is enough evidence to warrant a trial for the crimes charged.

The Prosecution

This is a good time to talk about who will be heading up the prosecution team and who will be looking after the defendant. All criminal cases are prosecuted on behalf of the people by the district attorney's office; in fact, when the prosecutors refer to themselves in court, they call themselves as "the people." You will undoubtedly be dealing with a deputy or assistant, and you may deal with several (fig. 4–3). It is not unusual in large jurisdictions to have three or four different prosecutors assigned to and reassigned from your arson case. Sometimes it will be necessary to bring one prosecutor up to speed on a case in the morning and a different one in the afternoon. While this can be frustrating, it is a common practice in large jurisdictions, and it is something an arson investigator needs to learn to deal with.

Fig. 4–3. It will become your responsibility to assist the deputy district attorney with the prosecution of the criminal case.

The Defense

Every defendant in the United States has the right to counsel, and if that defendant can't afford to hire a lawyer of his or her own choosing, then one will be appointed for him or her at no expense (fig. 4–4); hence the **public defender**. In some jurisdictions there are government offices established that employ public defenders. In other jurisdictions the court will require and pay private attorneys to act as public defenders. Like private attorneys and prosecutors, there are some good ones, and there are those who are not as good or as committed as they should be. Defendants who can afford to retain and pay a private attorney for their defense do not receive a public defender.

Fig. 4–4. The defendant has a right to counsel and the best defense possible. The defense attorney may be a private attorney or a court-appointed lawyer.

A third manner in which a defendant may defend him- or herself from criminal charges is to declare to be **pro per.** In some states you may hear the term **pro se,** which means exactly the same thing. It means that is the defendant is going to act as his or her own attorney at trial without benefit of trained legal counsel. As you can imagine, this does not generally work out all that well for the defendant. The suicide doctor Jack Kevorkian found this out in 1999 when he decided to represent himself in a second degree murder trial. Dr. Kevorkian had been tried several times previously but had won every trial with the assistance of competent counsel. At his 1999 trial, he decided to declare himself pro se and was convicted and sent to prison. There is a saying that a lawyer who represents himself has a fool for a client, and this definitely applies to those of us who have not been to law school as well.

The Preliminary Hearing

A preliminary hearing, often referred to as a "prelim," is sometimes described as the trial before the trial, but its purpose is not to prove or disprove the guilt or innocence of a suspect. The entire purpose of a prelim is to give a short presentation of the heart of the case to a judge for him or her to decide if there is enough evidence to warrant sending the case to trial. There is no jury present during this procedure, and in fact, the defendant has no right to request a jury during this phase of his prosecution. The defendant may be present during the preliminary hearing, may have counsel, and may mount an affirmative defense, but the final decision will rest with the judge. The prosecution may call witnesses to the stand and may introduce physical evidence. The defense may cross-examine and dispute items introduced as evidence in an attempt to have the case dismissed. Most prosecutors will only do as much as they believe necessary to cause the defendant to be held to answer, as they wisely do not wish to tip their strategy before trial. Most defense attorneys do not put on an affirmative defense for the same reason. Keep in mind that the burden of proof at this point is much lower than at trial, and the object is to move on to trial if warranted. The judge has only one thing to decide: Is there enough evidence to convince a jury that the defendant committed the crime with which he or she is charged? If the judge believes this to be the case, he or she will order the defendant to be **held to answer**, and the case will proceed toward an eventual trial.

The Grand Jury

The grand jury system is an entirely different animal, and it's easy to see why grand juries are not generally used in large, busy jurisdictions. In fact, only about one-half of the states in the United States use a grand jury system anymore. It is time-consuming, it is cumbersome, and many in the legal community do not believe it is a particularly fair system of charging someone with a crime. There is an old saying in the legal community that "you can indict a ham sandwich." The point is that the burden of proof is so low that virtually anyone (or in the case of the ham sandwich, anything) could be indicted before a grand jury. To fully understand why grand juries are viewed in this light, you have to understand the following things: Grand jurors are selected from various databases just as regular jurors are. However, there is no screening process, and no one is questioned about bias or other normal disqualifiers. The prosecutor decides what evidence is going to be presented to a grand jury and what witnesses are going to be called. There is no judge present, and grand jury proceedings are secret. The target of the grand jury has no right to testify and cannot compel the grand jury to hear certain witnesses or testimony. After hearing the testimony presented by the prosecutor, the grand jury will then decide whether or not to deliver a **true bill** and indict the target of the grand jury, sending him or her to trial at that point. No jeopardy attaches in a grand jury proceeding, and if the target is not indicted, the prosecutor has the option to file criminal charges against him or her, or take the case before a different grand jury.

Plea Bargains

After a defendant is held to answer by the judge in a preliminary hearing or is issued a true bill by a grand jury, there is generally a considerable amount of time before a case will come to trial. It could be as little as six months, or it could be several years on some very serious crimes. There are a number of reasons for this, but generally defendants will waive their right to a speedy trial (on the advice of counsel) and delay going to trial for as long as possible. This is particularly true if they happen to be out on bail while awaiting trial or even if they are in custody awaiting a trial that could result in them being confined the rest of their lives. Witnesses and victims die or lose interest, their memories fade, and they become more difficult to find. Fire investigators retire and move out of state or get reassigned, and their memories of the circumstances surrounding the case also become less vivid. Reports and evidence get misplaced or lost, and computer hard drives crash. There are many good reasons for defendants to wait as long as possible to go to trial, particularly if they believe they are going to lose. During this period of time, it is not rare for the prosecutor and the defense attorney to talk and often strike a deal known as a **plea bargain**. Plea bargains are very much maligned and have a negative connotation to the general public, but there are several good reasons for them (fig. 4–5). In many ways they are the mythical win–win situation we all like to talk about. In most plea bargains, the prosecutor and the defense attorney try to find some middle ground where the defendant can plead guilty to a lesser charge or plead guilty to the original charge in exchange for a reduced sentence. Plea bargains normally have to be reviewed and accepted by a judge, and the defendant has to admit his or her guilt in open court. The prosecutor gets a sure guilty verdict and therefore a win in the case. This sure win is a very good thing. Anyone who has spent any time at all in the criminal justice system will tell you there is no such thing as a guaranteed outcome in a criminal trial. There is no such thing as an airtight case where the eventual verdict is a certain, predetermined outcome. You really never know what a jury is going to do, and you can never tell how things are going to go for the prosecution or the defense. If you doubt this, think back to the O.J. Simpson murder trial—not very many prosecutors in Los Angeles saw that one coming. More recently, one can point to the murder trial of Casey Anthony in Orange County, Florida. Like the Simpson trial, this one was covered in great detail by the media. The trial lasted six weeks, and on July 5, 2011, Casey Anthony was found not guilty of first degree murder, aggravated manslaughter of a child, and aggravated child abuse. It would be safe to say that very few people thought this would be the outcome of the trial. There was a great deal of public outcry, as there was when O.J. Simpson was found not guilty. Sometimes the plea bargain is the smart, sure, and cheap way to settle a case.

Fig. 4–5. The prosecutor and the defense attorney will normally have a pretrial meeting for the purpose of reaching a plea agreement if possible.

There are, of course, some cases that are much stronger than others, but there is no sure deal. On the other side of the equation, with a plea bargain, the defendant and his or her attorney have possibly avoided a felony conviction or saved the defendant some jail time, and the judge has cleared the case from the court calendar and made room for other cases. It's also a good deal for the citizens who have now saved a considerable amount of tax dollars and have put a bad guy away for a period of time in most cases.

Pretrial Hearings

When the prosecutor and the defense have not been able to come to some sort of a plea agreement, the case will go to trial. Who are all the players and what steps are going to be taken before the defendant actually gets to court? At some point there will be a **pretrial hearing**. This is a rather informal meeting that will include the prosecutor, the defense attorney, and the judge. The fire or arson investigator will not be included in the meeting, nor will the defendant. The judge will meet with the two sides and try to get an inkling as to how the trial might go and to possibly suggest some compromises that might save the taxpayers and the courts some time and money. In other words, the judge is going to see if there isn't some sort of plea bargain the two sides haven't thought of or haven't been able to agree upon. Up to this point, the defense attorney and the prosecutor may have sparred with each other and told each other how strong their cases were in an attempt to bolster their position and therefore obtain a plea agreement that is favorable to one side or the other. Sometimes this works and sometimes it does not. However, once the two attorneys hear the judge's opinion of how he or she believes things are going to go for either side, positions are often changed, and an agreement is often reached. Statistics vary from jurisdiction to jurisdiction, but it is generally accepted that only about 5–10% of all criminal filings actually end up going to trial.

The Judge

But this case is going to trial, so we need to understand who all the players are and their basic jobs and powers. We will start with the **judge** and his or her responsibilities and powers (fig. 4–6). It is very important you understand who this person is and what he or she can do to your case. Ideally, the prosecutor will understand this, but he or she might not impart that knowledge to you, and you really need to know, particularly if you are new to the judicial surroundings. Although it is not legally required, virtually all judges are attorneys. Depending on the jurisdiction and the type of judge they are, they are normally either elected into office or appointed to the position and then confirmed by some sort of governing body. As already stated, judges are by far the most powerful persons involved in any legal proceeding, and you should never forget that.

Fig. 4–6. Once the case goes to trial, the judge becomes the gatekeeper of the entire legal process. The judge's word is literally law in the court, and the wise investigator keeps this in mind at all times.

Judges are like anyone else in many ways. They may be politically liberal or conservative, mild-mannered or harsh; they all have their own sets of values and beliefs. The difference is, they don't have to leave these opinions and feelings at home. Judges have the power to decide that laws are unconstitutional, and they have the power to declare laws to be invalid, at least as applied to a particular case. They can also summarily dismiss criminal charges against any defendant, even if you and the prosecutor believe the case is a sure thing. These decisions can be appealed and reversed, of course, but appeals are expensive and not many DAs have the resources available to appeal a case once a judge has dismissed it.

There is a pretty good reason for this if you give it some thought. Consider a large jurisdiction that might run dozens of criminal trials in a single day. It is large and impersonal if you are an outsider, but within their system, the judges, the prosecutors, and the DA all know each other, and they all know each other's feelings and political leanings.

After a while, investigators will come to know which judge will sign a search warrant with little scrutiny and which judges they want to avoid at all costs. Investigators who file arson cases on a regular basis, even in the largest jurisdictions, learn which prosecutors will really go after an arsonist and which ones will just go through the motions. No prosecutor wishes to alienate a judge—or maybe all the judges—by going through a lengthy and expensive appeals process and then face that judge on a nearly daily basis for the rest of their career in that venue. When a judge issues a poor decision, and they all do from time to time, it is best just to shrug it off and get on with the next case.

As an example, here is a personal experience when I had a criminal case that was dismissed when it should not have been. While investigating a small fire that had been set in an apartment building, my partner and I were approached by one of the residents, who told us we should speak with a particular woman who lived in the building. The neighbor stated that he believed the woman had at one time been in jail for arson, and he had seen her near the area of origin just before the fire was reported. Both of these pieces of information were extremely important to us, and of course we went to talk to the woman in a very timely manner. Keep in mind that while this is interesting information, neither statement made by the neighbor would be enough probable cause to arrest the woman in question. I knocked on the door of the apartment occupied by the woman, and she answered. I identified myself, I told her whom I worked for and the purpose of my visit, and she invited my partner and me into her apartment. After asking the woman her name and looking at her identification, I asked her if she had been home at the time of the fire and she responded, "Yes, I set the fire." In the arson investigation business, this is known as a clue. It is also known as a spontaneous statement; the woman was not under arrest or in a custodial situation, and therefore *Miranda* had not attached, at least not in my opinion or that of the prosecutor. I then placed the woman under arrest, advised her of her Miranda rights, took her to the jail, and booked her. After taking her to an interview room, she informed me that she no longer wished to speak with us and exercised her right to counsel, or as we say in the investigation business, she "lawyered up." I wasn't too concerned about this because we had her confession at the scene of the fire, and we had developed other evidence as a result of that confession. This was, at least in my mind, one of those slam-dunk cases that you hear so much about.

When it came time for the preliminary hearing, however, we drew a judge who was not known for her sympathy toward law enforcement. During my time on the stand, I was asked if I had read the defendant her Miranda rights as soon as she answered the door at the scene of the fire. I answered that I had not, as she was not in custody at that time, and I actually had not asked her any questions beyond identifying her and finding out if she was even home when the fire occurred. On the motion of the defense, the judge dismissed the case, stating that the defendant should have been Mirandized the moment she answered the door, as we had knowledge of her previous conviction for arson.

This is a ridiculous application of the *Miranda* decision, and not one judge in a thousand would have made such a ruling. However, the story is illustrative of the power and discretion that judges have and exercise. The case was never refiled, and the defendant was never prosecuted for that fire. The DA didn't even consider appealing the decision, even though he said that he thought the judge's interpretation of *Miranda* was flawed. He told us that he faces this judge on a regular basis and did not wish to alienate someone so powerful that he works with regularly. In retrospect, I can't say I blame him for that.

Most jurisdictions in the United States have some procedures for oversight into the actions on judges. However, the reasons for the oversight do not generally get into the merits of the decisions made by judges. Those can only be contested in an appellate process. Judges are removed from office for such offenses as taking bribes, open defiance of a binding court order, ruling where a clear conflict of interest exists, embezzlement of court funds, or a conviction of a serious offense unrelated to the judicial services. Judges in the United States also enjoy complete and absolute immunity for personal liability in the form of monetary damages from their discretionary judicial acts; they are indeed never held responsible for their decisions.

Judges should always be addressed as "Your Honor" or "Judge," and it never hurts to insert a few "Sirs" or "Ma'ams" at the appropriate times. You should never approach the **bench** (the judge's physical location in the court) without permission, and you should never speak to a judge without first asking permission to do so. You should never argue with a judge, and if you are ordered to do something by the bench, you should do so immediately and to the best of your ability. Anyone who ventures into a courtroom should always remember the judge is absolutely and always 100% in charge of that court. If you keep this in mind, your time in court will be much easier and more productive.

The Bailiff

Second only to the judge in authority in the courtroom is the **bailiff** (fig. 4–7). The bailiff is a court official and also usually enjoys the same status as that of a peace officer. They generally wear a uniform and carry a firearm. They have full powers of arrest, and in many jurisdictions they are deputy sheriffs or marshals. Their main job is to maintain order and safety in the courtroom and to help the court run smoothly by bringing in witnesses, escorting them back out of the courtroom, and in criminal trials, taking charge of defendants who are in custody during court proceedings. Judges depend heavily on bailiffs to enforce the orders they issue in the court, and generally you can consider the bailiff to be acting on the judge's behalf.

Fig. 4–7. The bailiff is there to keep order in the court and to ensure the safety of all of the participants in any kind of trial. Bailiffs are usually sworn police officers, deputy sheriffs, or U.S. marshals depending on the venue.

The Court Reporter

There is generally a **court reporter** present who is recording every word spoken while the court is in session (fig. 4–8). When the court is not in session or a sidebar has been requested, the conversation is usually not recorded. However, a judge can choose to have the defense and prosecution in his or her chambers with the reporter and then the court can pretty much be considered to be in session, and what is said will be part of the permanent record. Choose your words very carefully when you are in court, because they will never go away.

Fig. 4–8. Virtually every word spoken in court is memorialized by the court reporter, and a permanent record is available to anyone who needs it.

Types of Witnesses

There are essentially two types of witnesses in a court proceeding, and you could easily end up being either one. The first category, the "factual" witness, is the most common. A **factual witness** is one who is there to testify to things that actually occurred during the incident, or at least they perceived them to have happened. A factual witness will tell you about something they saw, heard, smelled, or tasted. Factual witnesses are not allowed to venture an opinion as to what anything they saw or heard meant, and when they asked to do so, it generally brings a strong objection from the opposition.

The other type of witness is called an **expert witness**. An expert witness is one who is perceived to possess a special knowledge or training that would qualify him or her to venture an opinion in a trial. In fact, the only purpose for introducing an expert witness is to get his or her opinion. Both sides do this, and you would be surprised at how radically opinions differ. It is entirely up to the judge to decide who is an expert witness and who is not. You may very well be subjected to an examination of your qualifications from the judge or the defense attorney. The fact that you may have qualified 20 times before as an expert is meaningless if your credentials are challenged. There are ways you can prepare for this, which will be discussed in chapter 8.

The Investigating Officer

In many states there is a special status granted to one partic- ipant in the courtroom, and you might very well be that person. The prosecutor has the right to declare one person to be the **investigating officer**. This is usually the person who initiated the investigation of the crime; interviewed the witnesses, victim, and suspect; and wrote the report that set the criminal filing in motion. The investigating officer's job is to sit at the table with the prosecutor throughout the trial and function as his or her assistant in prosecuting the case. Once this status is granted, the investigating officer cannot be excluded from the courtroom by the defense attorney, and he or she becomes the solid and unbroken link throughout the trial. The investigating officer is expected to retrieve reports, find witnesses, and get them to court, as well as any number of tasks that might become necessary during the trial. This is a significant and time-consuming job. If you are going to be an arson investigator, this job is inevitably going to become yours at some point. Good trial preparation is the key to being successful in this job.

Summary

Courtrooms and their procedures can be intimidating to anyone who is unfamiliar with the court's proceedings and protocols. It becomes clear when you understand who each of the participants are in a trial and what their specific functions are.

There are two basic types of trials: criminal and civil. Understanding the purpose, limitations, and consequences of these two systems is essential to the arson or fire investigator.

There is a specific set of steps that must be covered in a specific order for court proceedings to advance in a timely manner and to ensure that we do not deprive anyone of all of their rights. Failure to observe these rights and protections can not only jeopardize your case, but you may also deprive your victims of their chance at receiving justice.

Key Terms

Arraignment. Arraignment is usually a criminal defendant's first appearance in court or before a judge on a criminal charge. At arraignment, the charges against the defendant will be read or the defendant will be asked if he or she is aware of the charges against him or her and will be asked how he or she wishes to plead.

Bailiff. The court official or a law enforcement officer who keeps order in the courtroom and handles various tasks for the judge and clerk such as calling cases to approach the bench. The bailiff is usually a deputy sheriff, marshal, or police officer.

Bench. The area occupied by the judge(s) in a court. All the judges together or collectively are called the bench.

Civil court. A court venue related to the rights of private individuals and legal proceedings concerning these rights, as distinguished from criminal, military, or international regulations or proceedings.

Contempt of court. There are essentially two types of contempt: (1) being rude or disrespectful to the judge or other attorneys or causing a disturbance in the courtroom, particularly after being warned by the judge; and (2) willful failure to obey an order of the court. The latter can include failure to pay child support or alimony. The court's power to punish for contempt (called "citing" one for contempt) includes fines and/or jail time (called "imposing sanctions"). Incarceration is generally just a threat and, if imposed, usually brief. Since the judge has discretion to control the courtroom, contempt citations are generally not appealable unless the amount of fine or jail time is excessive. Criminal contempt involves contempt with the aim of obstruction of justice, such as threatening a judge or witness or disobeying an order to produce evidence.

Court reporter. Court reporters typically make verbatim reports of speeches, conversations, legal proceedings, meetings, and other events when written accounts of spoken words are necessary for correspondence, records, or legal proof.

Deputy district attorney. A deputy district attorney is usually hired by the district attorney to discharge the duties of the district attorney (see below) under his or her direct supervision.

District attorney. A district attorney is an elected or appointed public official of a county or designated district whose duties are governed by state law. Generally, the duties of a district attorney are to manage the prosecutor's office, investigate alleged crimes in cooperation with law enforcement, and file criminal charges or bringing evidence before the grand jury.

Double jeopardy. Placing someone on trial a second time for an offense for which he/she has been previously acquitted, even when new incriminating evidence has been unearthed. This is specifically prohibited by the Fifth Amendment to the U.S. Constitution.

Expert witness. A person who is a specialist in a subject, often technical, who may present his or her expert opinion without having been a witness to any occurrence relating to the lawsuit or criminal case. It is an exception to the rule against giving an opinion in trial, provided that the expert is qualified by evidence of his or her expertise, training, and special knowledge

Factual witness. An individual who knows facts about the case. For example, a witness to a murder knows facts about the case. Factual witnesses saw things, and they know things that no one else has seen or knows. They have to give this information to those in the courtroom in order for it to be admitted as a type of evidence for the case.

Felony. A crime sufficiently serious to be punishable by death or a term in state or federal prison, as distinguished from a misdemeanor, which is only punishable by confinement to county or local jail and/or a fine. Felonies are sometimes referred to as "high crimes" as described in the U.S. Constitution.

Grand jury. A panel of citizens that is convened by a court to decide whether it is appropriate for the government to indict (proceed with a prosecution against) someone suspected of a crime.

Guilty plea. A statement by someone accused of a crime that he or she committed the offense.

Held to answer. A term that means the prosecutor will file charges against the alleged suspect, and the defendant will plead guilty or not guilty.

Infraction. The breach of a law or agreement; the violation of a compact.

Investigating officer. The investigator who is in charge of the investigation from the beginning of the incident until the charging of a suspect in a criminal trial. The investigating officer assists the district attorney in the prosecution of the trial and usually cannot be excluded during the trial.

Judge. An official with the authority and responsibility to preside in a court, try lawsuits, and make legal rulings. Judges are almost always attorneys. In some states, "justices of the peace" need only pass a test, and federal and state "administrative law judges" may be lawyer or nonlawyer hearing officers who specialize in the subject matter upon which they are asked to rule.

Misdemeanor. A lesser crime punishable by a fine and/or county jail time for up to one year. Misdemeanors are distinguished from felonies, which can be punished by a state prison term.

Nolo contendere. Latin for "I will not contest" the charges, which is a plea made by a defendant of a criminal charge, allowing the judge to then find him or her guilty, often called a "plea of no contest."

Not guilty. The plea of a person who claims not to have committed the crime of which he or she is accused, made in court when arraigned or at a later time set by the court.

Plea bargain. A negotiated agreement between a criminal defendant and a prosecutor in which the defendant agrees to plead guilty or no contest to some crimes, along with possible conditions in return for reduction of the severity of the charges, dismissal of some of the charges, or some other benefit to the defendant.

Preliminary hearing. A preliminary hearing is used to determine whether probable cause exists to believe that the offense charged in the indictment has been committed by the defendant.

Pro per. *In pro per* is a short form of the Latin phrase, *in propria persona,* or "in the person of yourself." The full term, *in propria persona,* is hardly ever used in court. A person who is acting *in pro per* is called a pro per. In court, pro per and pro se are equivalent.

Pro se. *Pro se* is Latin for "for self" or in one's own behalf. You appear pro se in a legal action when you represent yourself directly in a legal action (in or out of court) and do not have an attorney speaking or writing for you. Pro se refers to representing yourself in any type of legal matter without the benefit of legal counsel.

Pretrial hearing. A meeting in which the opposing attorneys confer, ordinarily with a judge, to work toward the disposition of a case. In such meetings the discussion is related to the matters of evidence and narrowing of issues that will be tried.

Public defender. An elected or appointed public official who is an attorney regularly assigned by the courts to defend people accused of crimes who cannot afford a private attorney.

Tort. Civil wrongs, as opposed to criminal offenses, for which there is a legal remedy for harm caused. Torts fall into three general categories: intentional torts, negligent torts, and strict liability torts.

True bill. A true bill of indictment is the agreement of a grand jury that probable cause exists to order a defendant to stand trial on the charges in the indictment. When this occurs, the grand jury is said to have indicted the defendant. The defendant can then be brought to trial. If the grand jury returns a "not a true bill," that means they were unable to find probable cause needed to indict the accused.

Review Questions

1. What are the responsibilities of the district attorney and the public defender in a criminal trial?
2. What is the major difference between a criminal case and a civil case?
3. What is the difference between a preliminary hearing and a grand jury?
4. What is the entire goal of both preliminary hearings and grand juries?
5. What happens at an arraignment?
6. What are the possible penalties upon conviction of a felony crime?

Discussion Questions

1. Discuss the benefits of a plea bargain to both the defendant and the prosecution in a criminal case. Include the benefits to the taxpayers in the discussion.
2. Discuss the advantages and pitfalls of a defendant functioning in a pro per or pro se role during his or her criminal trial.
3. Discuss why some jurisdictions might choose to use the preliminary hearing option in a criminal case, while other jurisdictions might choose to convene a grand jury.

Activities

1. Spend a day in court watching a criminal trial and make note of how the proceedings occur. Then spend a day in court watching a civil trial and note the differences and the similarities.

2. Make an appointment with a deputy district attorney and interview him or her to find out what type of information DAs look for when they are filing a criminal case. Ask what type of red flags they look for before rejecting a filing and then ask what percentage of filings they reject and for what purposes.

5

Preparing for Trial

Learning Objectives

Upon completion of this chapter, you should be able to:
- Describe the way juries are selected and the importance to your case.
- Explain what you as an arson investigator can do to assist the district attorney in his or her efforts to try a suspect.
- Describe the role of the investigating officer in an arson case.

Case Study

Once the jury is seated, there are a couple of things that you should be aware of for your own peace of mind and to avoid some embarrassing situations. The first thing is that during your trial, no parties involved in any way with the trial are allowed to talk to the jurors for any reason whatsoever. That means no lawyers, witnesses, or anyone else is to speak to the jurors outside the courtroom for any reason at all. For example, I was once involved in an arson trial in Southern California in which a man had set his own house on fire and then told one of the arriving firefighters what he had done. This is known as a spontaneous statement and therefore outside of the *Miranda* requirements. In arson investigation circles this is also known as a clue and considered to be a very good thing for proving your case. This firefighter was a young, single man in his mid-20s, and while on the witness stand he noticed that one of the jurors was a very attractive young, single woman in her mid-20s. During the lunch break that day, he sought her out in the courthouse cafeteria and sat down with her for some pleasant conversation and courthouse food. The juror told him she didn't think they should be eating and talking together, but that did not deter our young firefighter. As soon as lunch was over, she reported the incident to the bailiff, who, of course, informed the judge. We spent the rest of that day and most of the next one trying to stave off a **mistrial.** In fact, we were very lucky because what this firefighter did could have constituted a reversible error and thus could have potentially been used to overturn our eventual conviction. No one had explained to the firefighter how inappropriate his actions were, and this was the first time he had ever testified in a criminal trial. I made it a point from then until the end of my career to explain very carefully to every potential witness from my department what was expected of them and what they were expected not to do. While some investigators may feel that the prosecutor has the responsibility to prepare all the witnesses for their courtroom appearances, I've always felt that if I could help in some way, I should. This is one area where you might have a good deal more influence with members of your fire department than the district attorney would. Most firefighters go through their entire career without ever ending up in court. Arson investigators are going to be there on a regular basis, so I believe it is incumbent upon them to prepare their members for testimony.

Introduction

By now you should have a pretty good understanding of who the players are and what your role might end up being. Sooner or later, if you work any cases, you are going to end up going to trial. So now let's get you prepared and make sure you understand what to expect.

As discussed in chapter 4, there are two basic types of trials. There are civil trials where one party brings legal action against another and seeks to gain a financial settlement for being hurt in some manner. Parties will often bring suits against third-party companies for faulty products that they believe started the fire or harmed someone in some manner. The purpose of these torts is to relieve one party of liability by placing that liability on the third party. These suits are known as **subrogation.** For example, subrogation occurs when an insurance company that has paid off its injured claimant takes the legal rights the claimant has against a third party that caused the injury and sues that third party. If you do more fire investigation than arson investigation, or if you work as an investigator for a private company, you will eventually become involved in a civil case. They are exclusively about money and never involve anyone going to jail directly. However, information derived from one of these civil cases could easily become a criminal matter.

For the purposes of preparation, we will assume that we are going to be involved in a criminal trial with the expressed purpose of putting a guilty person in jail for committing a crime. When defendants decide that they want their day in court, there are two routes they can take, and the choice is entirely theirs. A defendant can decide to have a **trial by jury** or a trial to be decided by the judge with no jury, also known as a **bench trial**. This term is derived from the place where the judge sits, which is also known as the bench. There is an old saying among prosecutors and law enforcement that guilty people will demand a jury trial, while innocent people will ask for a bench trial. While the reasons for this might not seem clear on the surface, there are reasons this philosophy might be valid. No one can possibly understand what a jury of 12 people might be thinking; no one can know what is going to be important to them and what is not; and no one can predict with any real accuracy what they are going to do once they get into the jury room (fig. 5–1).

Fig. 5–1. All defendants have the right to a jury trial. Here a group of citizens have been empaneled as a petit jury.

The judge, on the other hand, is much more predictable. The judge understands the law because, after all, nearly all judges are lawyers. They usually know the prosecutors and defense attorneys, and they are able to separate the important parts of a case in a way juries cannot. They are much less impressed with clever attorneys who put on a good show and more concerned with discerning the facts of the case. This means the case generally takes less time, and most cases will end with a predictable outcome. That verdict and the sentencing might not be what you had hoped or believed what it should be, but that is an entirely different subject.

Trial preparation should begin when you are on your way to investigate the fire. Assume that every fire you ever look at is going to go to trial someday. Assume that every picture you take is going to be scrutinized by a judge or jury someday. Assume that everyone you talk to is going to become a key witness, and therefore you must carefully record what they say and put it into a clear and concise report. Assume these things, and you will find yourself well prepared when your assumptions do come true. No one can accurately predict which cases will end up in a criminal or civil trial, so it is incumbent upon you to prepare as if every fire investigation is going to end up in court. Approach every investigation with this attitude, and you will generally be well prepared when you do eventually end up in a courtroom.

This is a good time to talk about juries, jurors, and jury selection— and your part in it. You probably will have very little or nothing to do with the jury selection, and the prosecutor probably will never speak to you about it because you don't really have a need to know. However, ignorance of juries, how they are selected, and how they can impact you and your case could be devastating. The fact that you may never be directly involved in a jury selection does not release you from the obligation of understanding how a jury is selected and how it may impact your job.

A criminal or civil trial jury is referred to as a **petit jury**, which is simply a term to differentiate it from a grand jury or a bench trial. Serving on a jury is usually a compulsory duty rather than one that people volunteer to do. Jury pools are drawn from a variety of sources, but the most common ones are the rolls of registered voters and licensed drivers in a particular state. A little known fact is that a judge can send officers to any public place in the community such as a restaurant or a department store to bring people in to bolster a jury pool. This is a radical departure from the common method of stocking a jury pool, but as we have discussed, the judge is an incredibly powerful individual and has the right to do this.

Once a pool of prospective jurors is established, the defense attorney, the prosecutor, and the judge begin the task of **empaneling** or selecting a jury for the trial. Normally there are 12 jury members and 2 alternates, but this is not a constitutional requirement. Grand juries usually involve more than 12 jurors, and civil trials may have as few as 6. In fact, in 1970 the U.S. Supreme Court said in *Williams v. Florida* that "the 12 man panel is not a necessary ingredient of trial by jury." However, the 12-person jury is certainly the most common and the traditional number for a criminal trial (fig. 5–2).

Fig. 5–2. The prosecutor will be afforded the first opportunity to make an opening statement to the jury, as the burden of proof is on the shoulders of the people.

Being called to be a member of a jury pool is by no means a guarantee that one is going to sit on a jury during a trial. There are many reasons why most people are excused from jury duty. The most common reason prospective jurors are dismissed is that their job in some way precludes them from serving, or they are related to someone whose job precludes them. Typically, police officers, firefighters, doctors, politicians, lawyers, and several others whose occupations cause them to be involved in the criminal justice system are excluded. Husbands, wives, brothers, sisters, and children of members of these professions are also routinely dismissed, because either the prosecutor or defense attorney believe that they will be influenced by their relatives' professional experiences. Jurors are also dismissed on the grounds of medical hardship, financial hardship, or extreme inconvenience.

Once the pool has been narrowed to a group of people with no real ties to anyone in the above professions or any personal problems whatsoever, the attorneys go to work looking for the most favorable jury for their purposes. This has become such a big deal now that an Internet search on "jury selection consultants" will yield a host of companies nationwide made up of lawyers, psychiatrists, body language experts, and social psychologists, who will help either side select the most favorable jury possible. While this tactic is rarely employed by the district attorney's office, it is a common practice among high-profile defendants who can afford such a service. Both the people and the defendant will have the opportunity to examine the prospective jurors through a process known as **voir dire**, which can include questionnaires, general questions answered by a show of hands, or direct questions asked of individuals that require individual verbal answers (fig. 5–3).

Fig. 5–3. The defense attorney will also be afforded the opportunity to make opening statements and tell the jury what he or she will prove to them during the trial.

During the voir dire, each side will have a limited number of **peremptory challenges,** in which they do not need a reason to dismiss a juror, and an unlimited number of **challenges for cause**. It is during the challenges for cause that some citizens will deliberately exploit this to get out of jury service. If you raise your hand when you are asked if you believe a defendant is guilty just because he or she has been arrested and brought to trial, you will be dismissed. We would have a much better legal system if good citizens would do their civic duty and honestly serve when called upon for jury duty. If you do actually feel that defendant is guilty just because he or she has been arrested, you should certainly answer the question honestly. However, this ploy for avoiding jury service is well known, and many people know exactly what to say if they don't wish to serve on a jury.

In capital cases where the death penalty may come into play, juries are often required to be **death qualified**. In other words, if the people are seeking the death penalty, the jury must be made up of jurors who will impose death if the defendant is convicted and if they believe death to be the appropriate punishment. This practice has been ruled constitutional by the U.S. Supreme Court, although a number of groups obviously do not like the practice.

Every fire investigator should be aware of what **jury nullification** is. Essentially, jury nullification is a jury rendering a law useless by their decision. Jury nullification occurs when a jury returns a verdict of not guilty despite its belief that the defendant is guilty of the violation charged. The jury in effect nullifies a law that it believes is either immoral or wrongly applied to the defendant whose fate they are charged with deciding. It happens for several reasons and can be really frustrating to the prosecution team when a defendant is acquitted regardless of the weight of the evidence. An example of jury nullification might be a jury who sets an arsonist free because the arsonist was shown to have been abused as a child. In this case sympathy for the arsonist's past might cause the jury to ignore his or her criminal activities, which is the very definition of jury nullification. Jury nullification is not illegal, but judges have no obligation to inform the juries of the practice, and a juror can be removed from the case if the judge believes the juror has knowledge of the power of nullification. The key for you is to recognize that this type of thing could very well happen in a case you are involved in. There isn't much you can do about it, but you can prepare yourself emotionally for a bit of disappointment in the legal system from time to time. You are also going to run into situations that don't rise to nullification, but where a defendant is convicted but then given a very light sentence that might consist of only probation or a suspended sentence. You may very well convict an individual of arson who will never spend a day in jail or prison. This is a lot more common than outright nullification. Any arson investigator who has been to court a significant number of times can probably point to some cases in which the defendant was convicted but did not receive as severe a sentence as he or she deserved.

Part of the reason for this feeling is that the investigator has invested a good deal of his or her time and effort into examining the fire scene; interviewing witnesses, suspects, and victims; writing good reports; and then filing them with the district attorney. This isn't something that only takes an hour or two, but a lot of time, hard work, and effort. You begin to take ownership of the case, and in many instances, you form a bond of sorts with the victims of the crime. When you believe the punishment doesn't match the crime, you feel a little bit cheated, and some investigators might feel that in some way they failed the victims of the crime. These are disappointments you are going to have to learn to live with and not to become discouraged. Courtroom procedures are not always going to go the way you want them to.

When preparing for trial, particularly your first trial, one of the things you can do that will help the prosecutor is to develop a detailed and accurate **curriculum vitae (CV)**. A CV is basically a short resume of your fire department training, education, and experience in relation to fire and arson investigation. It varies from a full resume as the only items that should be included are those that are germane to the purpose of the CV, which is to help the prosecutor to know your strengths and areas of expertise. The idea is to cull the significant factors of your fire department career from the others and to save the prosecutor the time of going through your entire fire service career record looking for the meaningful parts. If you have an Associate of Science degree in Fire Science, that should be included, whereas an AS in Accounting would not be meaningful. Certificates such as **National Fire Protection Association (NFPA)** Fire Investigator would be significant, while a NFPA certification as a Public Information Officer would not (fig. 5–4).

The format is another important factor in your CV. There is no reason to print it on parchment or include pictures of yourself or to mention your grade point average in high school. You are not applying for a job or promotion; you are trying to give the prosecutor a list of the reasons why a judge should qualify you as an expert witness and a jury listen to your opinion. While most resumes are pretty much organized chronologically and categorically, most CVs are organized by the order of importance. What is really important can be argued, but the vast majority of jurors really like their expert witnesses to have some formal education. If you have an AS in Fire Science and a BS in Fire Engineering, those two qualifications should be on the top of your CV. If you have been a member of the fire department for 17 years and an arson investigator for 8 years, that should come next. The remainder of the CV should contain specialized training received and certificates obtained. Try to keep the CV to one page, and remember that this is a living document. Keep it on your computer, and add new training on a regular basis. Keep several copies with you, and give one to each new prosecutor as they join the case or replace the last deputy district attorney. Remember that you are just one of dozens of investigators or police officers they work with each day, and they can't be expected to remember details about your career. Most prosecutors and attorneys will not ask you questions they don't already know the answers to; a complete and up-to-date CV will help prompt them to ask the right questions to enhance your qualifications in the eyes of jury.

CURRICULUM VITAE

ROBERT W. COMPTON

Captain, Long Beach Fire Department, 21 years of service

Commanding Officer of the Fire and Arson Unit, seven years

Certified Peace Officer, California Commission on Peace Officer Standards and Training (POST)

Associate in Science Degree in Fire Science 1991

Bachelor of Science Degree in Public Safety 1995

NFPA Fire Fighter I 1992

NFPA Fire Fighter II 1992

NFPA Fire Officer I 1994

NFPA Fire Investigator 2002

International Association of Arson Investigators Certified Fire Investigator

National Fire Academy Courses:

Fire/Arson Origin and Cause Investigations (R206) 1996

Fire Cause and Origin for Company Officers (R811) 1998

Forensic Evidence Collection (R214) 1999

Electrical Aspects of Fire Investigations (R255) 2001

Crime Scene Videographer, National Police Forensic Video Association

Post Blast Investigator School, Federal Bureau of Investigation

Advanced Post Blast Crime Scene School, Federal Bureau of Investigation

Drug Lab Hazardous Material Operations, Long Beach Police Department

---

Fig. 5–4. Arson and fire investigators should prepare a CV for the prosecutor that outlines any education, experience, and specialized knowledge they may possess.

For the purposes of this discussion, we will assume that you will be the investigating officer. This is probably the most important role an investigator can have at trial. The **investigating officer** is basically the prosecutor's right-hand person. The investigating officer has literally been in charge of the fire investigation from the onset of the incident and should be very familiar with every aspect of the case and with every potential witness who will testify. The defendant is being prosecuted as a result of the hard work of the investigating officer, so no one will be as intimately familiar with the details of the case as is the investigating officer. The court recognizes this position, and as a result, the investigating officer has the right and obligation to assist the people in the prosecution of the case.

As the investigating officer, you will sit at the table with the prosecutor and assist him or her in virtually every detail of the trial. You will not be excluded from the courtroom when other witnesses are, and you will act as the liaison between the district attorney's office and any number of other agencies including yours (fig. 5–5).

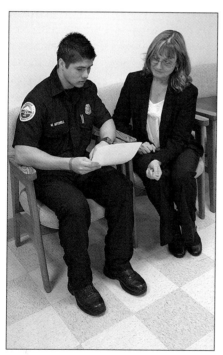

Fig. 5–5. It is your responsibility to make sure that the prosecutor understands everything on your CV that might become important during the trial. Spend some time going over your CV with the prosecutor before you go into the courtroom.

While there are many responsibilities that the investigating officer must assume, the number one responsibility is providing reports, pictures, and articles of evidence to the district attorney and providing them in a timely manner. You may work in a city of nearly 400,000 people with an ISO Class 1 fire department of nearly 500 paid members, and you still probably will not have a fire department evidence custodian or a records manager.

If your position as an arson investigator requires you to be a sworn peace officer, your reports might be filed with the police department rather than the fire department. There might also be a separate filing system for reports, pictures, laboratory analysis reports, and witness statements. As the investigating officer in an arson case, you will be responsible for providing the prosecutor with all follow-up reports that have been written after the arrest or charges were filed, providing photographs that may help in the prosecution of the case, and, of course, retrieving items of evidence at the proper time.

At some point before the trial begins and sometimes even after that, the defense will file a motion for discovery, and your department will receive what is called a **subpoena duces tecum**. *Duces tecum* is a Latin phrase that literally means "bring with you under penalty of punishment." A *duces tecum* is the defense team's way of exercising their right of discovery. In other words, they have a right to know everything about the case that you do, including any **exculpatory evidence** if it exists. This means that they get a copy of every report, picture, handwritten note, videotape, and anything else that is in the case file. Unless your department is are full of clerical staff, it will be up to you to copy the reports, pictures, and all of the other afore-mentioned items for the defense attorney. You can, of course, bill the other side to recover the costs of replicating all of these items, but you must provide them. We will discuss subpoenas and warrants in greater depth in chapter 6.

The very fact that the defense attorney and everyone else involved in the case will have access to these documents is an excellent reason to be concerned about accurate note taking at the scene of the fire and accurate report writing afterward. After serving the subpoena *duces tecum* the defense will examine your notes and compare them to your final report.

Some investigators will destroy their notes after writing their final report, but this practice is usually harder to explain at trial than an inconsistency in your report. When the defense attorney asks why you don't have any field notes or sketches, you are going to have to explain why you destroyed them. It is very difficult to come up with a really good reason for destroying your field notes, and doing so usually plants the thought in the minds of the jurors that you had something to hide. When jury members hear that a fire investigator destroyed his or her field notes, the first thing they will wonder is why and what might the investigator be hiding? Put yourself in the place of the jury members and the defense attorney, and you will probably have the same thoughts. Take good notes and write good reports, and save yourself this problem.

On the subject of notes, now that you know everything you write is going to be read by any number of people, it should occur to you that what you write is very important. Snide comments or personal observations about witnesses and suspects that fall outside of the scope of the investigation are never a good idea. Your field notes should be self-serving. I know that term has a negative connotation, but it is the absolute truth. The defense attorney is going to come at you from the standpoint of trying to make a jury believe there was something about the defendant that you didn't like and that is the reason you accused the defendant of arson. Make sure your notes are nothing but professional, and as you write them, assume that they will be read aloud in court. This doesn't mean you don't take fair and accurate notes. No one wants to put an innocent person in prison, but I was always just as concerned about letting a guilty person go free because of a mistake I made. Don't make that mistake in your notes.

Items of **evidence** are really critical in arson investigation cases, and the job of making sure they are in the hands of the prosecutor at the time of trial is the responsibility of the investigating officer. Most fire departments do not have their own evidence rooms complete with an **evidence custodian** to book items of evidence in and retrieve them, therefore assuring the courts of the **chain of custody** of the aforementioned evidence. An old wall locker sitting on the apparatus floor at a fire station that serves as your evidence locker is an extremely bad idea—a situation waiting for a disaster to occur to your case in court. Police departments do not keep evidence involved in felony cases in old wall lockers with a piece of wire routed through the handle. If they did, it would provide a perfect invitation for the defense attorney to bring up the subject of evidence tampering, evidence contamination, and evidence spoliation. And in this particular case, the defense attorney would be absolutely correct. Arson is a felony in every state in the union, and this wall locker on the apparatus floor is no place to store the hammer the defendant used to break the window to pour gasoline into the bedroom of the victim and start the fire. Sooner or later, someone in the fire station is going to need a hammer and think of that one. I don't think that anyone, even those of us who are not police officers, firefighters, or emergency medical technicians would think it is OK to keep a rifle used in a murder in an old wall locker in the parking garage of police headquarters. I don't think anyone would think it was OK for police officers to borrow the rifle to go deer hunting and then put it back before the murder trial. As ridiculous as this sounds, I have observed fire department investigators place arson evidence in an old wall locker on the apparatus floor and then secure the locking mechanism with a piece of a wire coat hanger. I have yet to have anyone satisfactorily explain the difference to me.

Think back to the first O.J. Simpson trial and all that was made of the allegedly sloppy way the Los Angeles Police Department handled the evidence in the trial. Police officers carried items around in the trunks of their cars, and the physical evidence—more specifically, the contamination of that evidence—became a major factor in the trial and undoubtedly helped yield what most people believe was an unjust verdict.

Use the police department's evidence room and their evidence techs if they will allow you to do so. If they won't cooperate with you, a visit with the city attorney and the chief of the fire department would be in order. Fire investigation and the prosecution of arson are serious endeavors, and evidence is a serious part of those endeavors.

Even if the police department is cooperative, there are some items that you will not be allowed to book into police evidence, and with valid reason. Most police departments will not book any flammable liquids, the containers those flammable liquids came in, lighters, matches, or similar items into evidence because of their hazardous nature. This is a very sound and reasonable policy for police departments to adopt as these items are by their nature dangerous. Of course, these items are often found at arson scenes. It is conceivable that you will have to establish some sort of storage room for these items. Most of the time a judge would never allow you to bring these items into a courtroom, but that does not absolve you of the responsibility of handling them properly. Most judges and jurors are reasonable people, and they will probably understand why you were not allowed to book a half-full container of gasoline into evidence in the police station. Still, they have a reasonable belief that these items should be stored and available for their examination if they choose to do so. I absolutely guarantee you that the defense attorney is going to inquire about the evidence at some point in the trial. You should consult with your city attorney and the district attorney's office on the best way to handle this.

You are the main connection to the case for all of your witnesses, victims, and fire department personnel involved in the prosecution case. One of the things that will prove helpful is to develop some type of a relationship with the victims and witnesses. Ideally, you already know your fellow firefighters pretty well, and they are a little more accessible to you. In most of your cases that do go to trial, there is probably a significant period of time between the arrest of the suspect and the beginning of the trial. Often times this period of time will exceed a year, and after that amount of time, witnesses and victims are sometimes very difficult to find if you have not kept in touch with them. This doesn't mean you have to take them fishing or invite them to the department picnic, but you should stop by their homes on a regular basis or call them on the telephone to check on their welfare and update them on the progress of the case. This accomplishes two things: It allows you to stay current with their most recent residence, and it elevates you to more than just the cop they talked to at the fire two years ago. Two weeks after a fire, most witnesses are chomping at the bit to testify and put the bad guy behind bars; two years after the fire, not so much. It is up to you to routinely update the witnesses and victims on the progress of the case and keep them interested. It will also help you in doing your job when the district attorney gives you a handful of subpoenas to deliver. It is much easier to find someone when they have moved a month ago than it is when they moved two years ago. All you really need to do is to stop by their place of residence every few months and give them a quick update on the case and a business card or two. You're going to find that you will meet family members and friends of the family who might very well help you find a witness that you really need in the future. Try to remember that while investigating fires and prosecuting people for arson is an everyday thing for you, it is probably a once-in-a-lifetime event for them. Connecting with witnesses and their family on a personal level really makes them want to cooperate and help you in your case. To paraphrase Jerry Maguire: Help them help you (fig. 5–6).

Fig. 5–6. You are going to be the main contact for all witnesses in the case. You will most likely be responsible for making sure they know they are due in court, and sometimes you may have to provide transportation.

Summary

Being prepared for your day in court is one of the most important things you can possibly do. Like anything else, careful preparation and forethought usually produce successful results. This begins with you taking good field notes and then using those field notes to produce a good report for the district attorney.

As an investigator, remember that every note you create, every picture you take, and every aspect of your investigation will eventually end up in the hands of the defense attorney through discovery. Every effort must be made to keep your report complete, accurate, and professional.

You should develop and maintain your CV to assist the district attorney in establishing your credentials as an investigator in a court of law. The CV should be updated on a regular basis and include all pertinent information that might assist the district attorney.

The position of investigating officer is extremely important and not a position to be taken lightly. You are the liaison for the district attorney to interact with witnesses, particularly those witnesses who are fire department employees. Make every effort to ensure that fire department employees are prepared for their court appearances and are familiar with the proceedings.

Introducing evidence and accounting for the chain of custody is a critical function that is often shared with the police department. Evidence preservation and storage is a special problem for the fire service due to the nature of some of the evidence.

No case is going to be successfully concluded unless you can find your witnesses and get them to show up to testify. You must, in many cases, develop a personal relationship with the witnesses in your case and develop a method of finding them when the time for the trial finally rolls around.

Key Terms

Bench trial. A trial conducted before a judge without a jury. In such trials, the judge decides both questions of facts and questions of law.

Chain of custody. A process that must be followed for evidence to be legally defensible (acceptable to courts and government agencies). It involves these main elements: The evidence collector properly identifies the evidence. The evidence collector must be a neutral party who has no personal interest in the test results The evidence collector tamper-proofs and secures the evidence at the collection site.

Challenge for cause. A challenge for cause is a request that a prospective juror be dismissed because there is a specific reason to believe that the person cannot be fair, unbiased, or capable of serving as a juror.

Curriculum vitae (CV). A CV includes a summary of your educational and academic background as well as teaching and research experience, publications, presentations, awards, honors, affiliations, and other details that pertain to your qualifications as a fire or arson investigator.

Death qualified. Refers to a jury dealing with criminal cases where death penalty is the most likely sentence. This type of jury consists of jurors who are not categorically opposed to the imposition of capital punishment and not of the belief that the death penalty must be imposed in all instances of capital murder.

Empanel. Select and seat the jury for a case. It is same as impanel. Empaneling is a jury selection process and involves many steps.

Evidence. Something that furnishes proof of a matter. In the legal context, it is something legally submitted in court or other decision-making body to ascertain the truth of a matter. Evidence may take various forms, such as oral testimony, videotape, and documents.

Evidence custodian. The person or persons duly appointed on competent orders as the custodian must be available to receive and release evidence and attend to other matters as required.

Exculpatory evidence. Evidence that favors a defendant in a criminal trial and tends to establish the defendant's innocence. It shows that a defendant had no criminal intent to commit the crime.

Investigating officer. The investigator who is in charge of the investigation from the beginning of the incident until the charging of a suspect in a criminal trial. The investigating officer assists the district attorney in the prosecution of the trial and is usually exempt from being excluded during the trial.

Jury nullification. Occurs when a jury returns a verdict of not guilty despite its belief that the defendant is guilty of the violation charged. The jury in effect nullifies a law that it believes is either immoral or wrongly applied to the defendant whose fate they are charged with deciding.

Mistrial. The termination of a trial before its natural conclusion because of a procedural error; statements by a witness, judge, or attorney that prejudice a jury; a deadlock by a jury without reaching a verdict after lengthy deliberation (a "hung" jury); or the failure to complete a trial within the time set by the court. A mistrial may be declared by the judge on his or her own initiative or upon the motion (request) of one of the parties to declare a mistrial.

National Fire Protection Association (NFPA). The mission of the international, nonprofit NFPA, established in 1896, is to reduce the worldwide burden of fire and other hazards on the quality of life by providing and advocating consensus codes and standards, research, training, and education.

Petit jury. An old-fashioned name for a jury that hears a civil lawsuit or criminal prosecution. *Petit* is a French word meaning "small," to distinguish it from a grand jury, which performs other duties, mainly to return an indictment or not. A person on a petit jury is part of the most common type of jury service.

Peremptory challenge. The right of the prosecution and the defense in a jury trial to have a juror dismissed before trial without stating a reason. This challenge is distinguished from a challenge for cause based on the potential juror admitting bias, acquaintanceship with one of the parties or their attorney, personal knowledge about the facts, or some other basis for believing he or she might not be impartial.

Subrogation. The substitution of one person in the place of another relating to a lawful claim or right. Subrogation commonly occurs in insurance matters, when an insurance company pays its insured client for injuries and losses, and then sues the party that the injured person contends caused the damages to him or her.

Subpoena duces tecum. A subpoena that requests items be brought with the person is called a *subpoena duces tecum*. A subpoena is an order directed to an individual commanding him or her to appear in court on a certain day to testify or produce documents for a pending trial.

Trial by jury. A jury trial is otherwise called a trial by jury. A jury is a group of law-abiding members of a community who have been assigned to arrive at an impartial decision on a legal issue. The jury may either arrive at a decision or make findings of fact that are then applied by a judge. It is thus different from a bench trial, where a judge or panel of judges makes all decisions.

Voir dire. *Voir dire* is a Latin term meaning "to see or speak." Voir dire is a legal procedure conducted before trial in which the attorneys and the judge question prospective jurors to determine if any juror is biased and/or cannot deal with the issues fairly, or if there is cause not to allow a juror to serve.

Review Questions

1. What industry might employ subrogation, and what might the goal of this type of lawsuit be?
2. What is the difference between a bench trial and a jury trial?
3. When should the fire or arson investigator begin preparing for trial?
4. What is a peremptory challenge during jury selection?
5. What is a challenge for cause during jury selection?
6. Who decides whether a witness is going to be designated an expert or not?
7. What is the process of questioning a prospective juror for fitness called?
8. Who does exculpatory evidence tend to favor?
9. What does the term *duces tecum* mean in English?
10. What is a curriculum vitae?

Discussion Questions

1. Discuss the possible reasons that a defendant would select a bench trial rather than a jury trial.
2. Discuss the reasons for taking accurate notes in the field and the reasons to keep them purely professional.
3. Define and discuss the role of the investigating officer.

6

Warrants and Subpoenas

Learning Objectives

Upon completion of this chapter, you should be able to:

- Describe the need for a subpoena and warrant process for the fire investigator.
- Define the difference between an arrest warrant and a search warrant.
- Describe the purpose of a *subpoena ad testificandum*.
- Describe the purpose of a *subpoena duces tecum*.
- Discuss what right *Terry v. Ohio* gives police officers.

Case Study

The last thing most firefighters think they are ever going to do is to end up serving subpoenas or swearing out arrest warrants. For the most part, this is true, but should your career take a turn toward the field of fire and arson investigation, you will find yourself with these new responsibilities. In this chapter we are going to learn a bit about subpoenas and warrants. We will discuss what they actually are, the different types, how you get them, and what to do with them. The world of subpoenas and warrants might be foreign territory to you, an intimidating one especially if you do not understand what is expected of you.

Until you actually have the experience of serving a subpoena or securing and serving an arrest warrant, there will be many questions in your mind. The purpose of this chapter is to give you some insight about this part of the criminal justice system. This chapter isn't designed to make you an expert, and there are procedural differences from state to state, but ideally you will know a good deal more about the subject by the time you finish the chapter.

Who Might Need a Warrant?

If your position as a fire or arson investigator bestows the title of **peace officer** on you, you will undoubtedly be involved in making an arrest sooner or later. If you work in a large metropolitan area, it will probably be sooner and on a fairly regular basis. The designation of peace officer grants authority to an individual beyond the power of making a citizen's arrest. In some jurisdictions peace officers might be known as "sworn personnel." This authority is usually outlined in the state criminal code, as are the qualifications of who can enjoy this status and what training they must complete. Once again, it varies from state to state, and there is no one-size-fits-all definition or description. Obviously, those who work for a police or sheriff's department fall into this category, and there are many more agencies and individuals who do as well. Some examples are agents of the liquor license commission, employees of a state regulatory commission dealing with gambling, or my personal favorite, firefighters who are assigned to the investigation unit and investigate crimes of arson and other criminal acts involving fire. Most people believe that the great advantage of being a peace officer is the authority to carry a firearm, and indeed most peace officers do just that. The real advantage, however, is the protection the title affords the officer when making an arrest. In the United States, a law enforcement officer can only make a lawful arrest in two circumstances.

The first circumstance is called **probable cause**. This is a legal term that means that through your investigation, you have uncovered enough facts and evidence to form an opinion that a particular suspect committed a crime and that he or she should be immediately arrested for it. There are several ways that you might arrive at probable cause, such as a neighbor reporting that he or she saw the person fleeing a burning house or the suspect making statements that would indicate that he or she set the fire. When you arrest someone, you will be asked by the booking sergeant what your probable cause is, and you will need to be able to state exactly why you thought you should arrest this person. Remember that probable cause doesn't come anywhere close to the level of proof needed to convict someone or to even charge that person with a crime. Many times, what looks like probable cause on Friday night may no longer look that way by Saturday afternoon, after a more exhaustive investigation. If that is the case, you turn your prisoner loose, and an apology of sorts might be a nice touch, although it is not required. As a law enforcement officer, you are shielded from false arrest or false imprisonment suits by the doctrine of probable cause.

Arrest Warrant

The second way a law enforcement officer can make a **legal arrest** is by the use of an arrest warrant. About 90% of all arrests are made on a probable cause basis, but there are some really good reasons for seeking an **arrest warrant**. Probably the best is when you have plenty of probable cause to arrest a suspect, but you have absolutely no idea where he or she is. It is not uncommon for criminals to vacate a particular geographic region after committing a crime and to lie low in another part of the state or even another state. Law enforcement has a wonderful computer system, and the term "small world" really does apply in this case. Once you obtain an arrest warrant, you can run this arrest warrant in that system, and if your arsonist from California is stopped for a traffic ticket in Arizona, he or she will be arrested and you will be afforded the opportunity to retrieve him or her. Felony arrest warrants and no-bail arrest warrants are taken very seriously by law enforcement everywhere, and they are happy to take your offender into custody.

There are cases where your probable cause might not be as strong as it could be, and you are a little concerned about arresting someone based on the evidence you have. Sometimes it is a good idea to take your report to a judge for an arrest warrant, and let the most important person involved make the decision to arrest the suspect or not. After a period of time, you may very well come to know a judge or judges well enough to informally speak with them about your case and see what advice they are willing to give you.

Laws vary from state to state, but generally an arrest warrant is needed to arrest an individual for a misdemeanor crime when it was not committed in the presence of the law enforcement officer, and to arrest a suspect for a felony in his or her home if it is not an emergency situation. Once again, if the suspect is out walking around in public, you would be perfectly within the law to make a felony arrest based on probable cause.

What It Takes

Now that you understand the reasoning behind an arrest warrant, let us discuss what it takes to achieve one. There are generally four requirements that must be met before a judge will grant an arrest warrant:

1. Above all else, you must be able to show enough probable cause to convince the judge that an arrest warrant is appropriate. This is generally accomplished by supplying the judge with a copy of your well-written and concise report. As the judge reads your report, the reason for the warrant should become apparent to him or her. The judge may have a question or two for you, but if your report is so poorly written the judge cannot understand what the probable cause is, your troubles are just beginning.

2. The warrant must be issued by a neutral or detached judge. In other words, the judge should be have no other knowledge of the case or the suspect and can take a fair and unbiased look at your report, and then decide if your investigation warrants an arrest.

3. Along those lines, your affidavit, which is based on your investigation report, must be free of any statements that you know to be untrue. In fact, you are going to be required to raise your right hand and swear an oath as if you were being sworn in for testimony at trial. Like your testimony at trial, this oath carries a penalty of perjury with it, which is, as far as I know, a felony in every state of the United States. Swearing out an arrest warrant is a very serious business, and judges tend to become upset if you lie to them. If you lie to them under oath, you are committing a crime, and they will make you pay for it. We will discuss what constitutes perjury

and what does not in chapter 7, but right now, knowing it is a felony is sufficient.

4. Finally, the arrest warrant must describe the person to be arrested in great detail. In chapter 2 we discussed the importance of knowing exactly whom you are talking to. This really comes into play when you are applying for an arrest warrant. A name and an address are not enough. This part of the arrest warrant must comply with the Fourth Amendment, and the term used is "particularly describe" the person to be arrested. Every detail that you learn about the person to be arrested by properly identifying your suspect is essential here. You need to do a good job of that.

These minimum requirements stem from the language used in the Fourth Amendment of the Constitution.

Bench Warrant

You have probably heard the term **bench warrant** at some point in your life. A bench warrant is a form of arrest warrant that is issued by judges and magistrates. Bench warrants authorize the immediate, on-sight arrest of the specific individual who is the subject of the bench warrant. The most common use of bench warrants is for persons deemed to be contempt of court. The most common form of contempt of court used in bench warrants is failure to appear at the appointed time and date for a mandatory court appearance. Indeed, the common street slang for this type of warrant is "failure to appear" (FTA). The bench warrant authorizes any law enforcement officer to take the subject into custody and to hold them without bail. The subject will then be brought before the judge, and the judge may set a bail amount or may hold the subject without bail if the judge considers him or her to be a flight risk or is just tired of the subject holding up the proceedings. Stall tactics are often used in court proceedings by defendants' lawyers in hopes of the prosecution losing interest in the case. Sometimes this tactic actually works as witnesses and victims do lose interest in the case, move to other cities, retire from jobs, or their memories fade.

Outstanding Warrants

An **outstanding arrest warrant** is one that has not yet been served. Across the United States there are literally millions of outstanding warrants. In some major cities, such as New York or Los Angeles, the number is more than a million. Still, the arrest warrant is one of the best avenues to eventually apprehend a suspect who is doing his or her best to avoid you. Some localities put some very serious restraints on those who have outstanding warrants, such as not allowing them to get a driver's license, apply for a passport, or even cross into Mexico or Canada for a day of shopping. Arrest warrants greatly enhance your chances of bringing a case to a successful conclusion. Do not avoid using them because you do not understand how to obtain one. Embrace the subject and learn how to be a more effective fire or arson investigator.

Search Warrant

By definition, a **search warrant** is a court-issued warrant that will allow you, the law enforcement officer, to enter and search specific areas for specific items. Before we discuss when a search warrant is needed, we should discuss when a search warrant is not needed. This may save you a good deal of time and effort. Under the Fourth Amendment to the U.S. Constitution, most searches by law enforcement require a warrant, but there are some exceptions, and you need to know about them.

Consent

Free and voluntary consent of the person in control of the property in question always yields a legal search, and any evidence found during that search will most certainly be allowed to be introduced as evidence. There are a couple of terms in the previous sentence that need to be discussed so we understand exactly what they mean.

Free and Voluntary

Free and voluntary mean just what they say. *Free* means free from coercion or threats by law enforcement or anyone else regarding what might happen if one does not allow the property to be searched. The search must be *voluntary*, meaning that the individuals are allowing the property to be searched without any coercion or threats. For example, telling the persons in control of the property that if they do not allow you to search their property, they will be arrested is probably going to be considered to be coercion. If they are coerced, they are probably no longer giving you their permission to search on a voluntary basis. It will probably be pretty easy for the defense attorney to make that argument. Once your search is declared to be illegal, so will the evidence you obtained from it.

However, it is perfectly acceptable and legal to explain to the persons in control of the property what is going to happen after they refuse to allow you to search the property or conduct an investigation. My explanation usually went something like this: "You have every right to refuse me access to your property and not allow me to search your property or conduct an investigation. However, it is my legal obligation to do those things, so I am going to have you and everyone else stay out of the house while I obtain a search warrant." This will probably cause the individuals to ask questions such as how long that will take and what they are supposed to do in the meantime. Be truthful and be professional, and above all else, be prepared to get that warrant if you say you are going to do it. Sometimes when the subjects realize that all they are doing is wasting time, they may change their minds and give you permission. Truthfully explaining what is going to happen next is never going to be considered coercion.

In Control of the Property

The second part of consent is the phrase "person in control of the property." On the surface, this might seem really simple to you, but it can be more complex and more confusing than you might initially believe. Rental properties bring up some interesting questions, and it is important to know what the phrase "in control of the property" means. Many landlords are under the impression that even though they have renters in a piece of their property, they are still the persons in control and therefore have the right to grant permission for the property to be searched. While this probably makes perfect sense to the landlord, and perhaps to most people, the courts do not always see things that way. As a rule of thumb, for traditional rental properties like houses and apartments, it is necessary to get the person actually living there to grant you permission to search the property. Landlords give up certain rights to the property they own when they rent it or lease it. They cannot, for example, enter the property while you are living there without your permission or without making certain arrangements. Another one of the rights they give up is the right to grant the authorities permission to search the property once it is rented.

Hotel and motel rooms fall into a slightly different category. Because of the transitory nature of the occupancy, control of the property is viewed a bit differently. In some jurisdictions, permission for searches of hotel and motel rooms can be granted by the hotel's management, without permission from the guests and without a warrant. It would be smart for you to check with the district attorney's office where you work and get their view on this question.

From time to time, the question comes up about whether the permission to search must be in a written form or if verbal permission will suffice. This tends to be a matter of departmental policy, and opinions vary greatly. While I have never seen any case law or legal requirement that the permission must be in writing, it certainly couldn't hurt.

Exigent Circumstances

Most people define **exigent** as meaning the same thing as emergency. That's close, but not exactly what it means. Exigent actually means urgent, pressing, or something requiring immediate attention. This could be very important in a court of law, for example, if a defense attorney were to try to alter the meaning of the word, or if we do not understand it ourselves. In the emergency services field, we usually think of an emergency as a situation where someone's life, or least their safety and well-being, is in jeopardy; this is not at all the definition of exigent. While we would all agree that a person screaming for help would certainly meet the definition of exigent, so would the need to pursue a fleeing felon into a structure where he or she has sought shelter. Exigent circumstances also describe a situation where evidence might be removed or destroyed if the authorities do not act in a swift and positive manner.

In the fire service, we are all about the exigent circumstances. I am a firm believer in respecting everyone's civil rights, but those civil rights also extend to fire investigators and how we do our job. We discussed *Michigan v. Tyler* in chapter 3, but now is a great time to bring it up again. I would never encourage anyone to attempt to usurp the U.S. Constitution, but I also don't believe that we should hamstring ourselves through ignorance. Now would be a wonderful time for us to read a key passage from *Michigan v. Tyler* again.

> We hold that entry to fight a fire requires no warrant, and that once in the building, officials may remain there for a reasonable period of time to investigate the cause of the blaze. Evidence of arson discovered in the course of such investigations is admissible at trial, but if the investigating officials find probable cause to believe that arson has occurred and require further access to gather evidence for a possible prosecution, they may obtain a warrant only upon a tradition showing of probable cause applicable to searches for evidence of a crime. We do not believe that a warrant was necessary for the early morning re-entries on January 22. As the fire was being extinguished, Chief See and his assistants began their investigation, but visibility was severely hindered by darkness, steam, and smoke. Thus they had departed at 4:00 am and returned shortly after daylight to continue their investigation. Little purpose would have been served by their remaining in the building, except to remove any doubt about the legality of the warrantless search and seizure later that same morning. Under these circumstances, we find that the morning entries were no more than an actual continuation of the first, and the lack of a warrant thus did not invalidate the resulting seizure of evidence.

Obtaining a Subpoena or Warrant

Now that we have covered the circumstances when a subpoena or warrant is not necessary, it's time to discuss those that do require warrants and subpoenas. You should never attempt avoid securing a search warrant; take advantage of every legal advantage afforded you. Don't make your job more difficult. Search warrants, like arrest warrants, are issued by judges or magistrates. They are court orders authorizing you to search a specific place for specific items. Generally you have to meet two criteria to even apply for a search warrant:

1. You must be able to show that a crime has been committed.
2. You must be able to show that a search warrant will lead to evidence of the crime and who committed the crime.

Search warrants are not fishing licenses, and judges take the issuance of such very seriously. After all, even with a search warrant, you are invading someone's expectation of privacy, even if that expectation is unrealistic. Search warrants are based on probable cause, which is in turn based on the totality of the situation. For example, let's assume that you are investigating a firebombing where the suspect constructed a **Molotov cocktail** using a very distinctive glass bottle that might have contained a beverage that traditionally comes in a six-pack. Let's say the wick was cut from a red, white, and blue beach towel, and you were able to see two people throwing a beach ball on the wick. Finally, let's say the victim of the firebombing was home and saw her ex-boyfriend speeding away in his 1978 Camaro Z28. In this case you would have plenty of probable cause to apply for a search warrant for both the suspect's house and the suspect's car. If I were writing the request for the warrant, I would be looking for empty or full bottles to match the one found at the scene, a beach towel with a piece missing that matched the description of the wick from the bottle, and any literature giving instructions on how to make such a device. I would also ask to seize any computers in the house so they could be examined for Internet searches on the construction of firebombs. These are all very reasonable requests, and it's likely that I would be granted the search warrant.

In Plain View

"In plain view" is a very interesting concept. *Michigan v. Tyler* is based on it, from an investigation standpoint. The court states that once the fire is extinguished, we can attempt to discover the cause of the fire, and any evidence found in plain view is admissible. It also states that any evidence found in plain view can be used for probable cause to obtain a search warrant. Let's take our scenario above, and let's say we are conducting a search of our firebomber's house, and we discover a woman chained to the bed in the suspect's bedroom. Obviously, this was not mentioned in our search warrant, but we have a pretty serious situation here. The law states that if an officer is legitimately on the premises (we are because we have a search warrant), and the officer's observation is from a legitimate vantage point (it is because the computer or any of the other evidence could be in the bedroom), and it is immediately obvious that the evidence is contraband (a woman tied up will probably meet this criteria), then the officer would be within his or her rights to seize the object in question. The same thing is going to apply to a pound of cocaine on the kitchen table or child pornography on the computer.

Return Warrant

After you seize any evidence, you must provide the resident of the property searched with a receipt. Then you must file a report with the court and tell it exactly what you seized when you executed the warrant. This is easy to forget, but not filing the report has a tendency to really upset judges. Learn to be methodical in your investigation practices.

Vehicles

Quite simply, there is a greatly diminished expectation of privacy in vehicles. In the United States the police do not need a search warrant to search a vehicle they stop on the road or in a nonresidential area. If they have probable cause, they can search the passenger area, the trunk, and all of the compartments in the vehicle. The Fourth Amendment states that people have a right to be secure in their persons, houses, papers, and effects. It does not mention automobiles or vehicles anywhere in the entire amendment. So how could this end up involving you, the fire investigator? Let's say Officer Muldoon stops a car because it is weaving and crosses into the lane next to it without signaling. I think you would be surprised how often that happens! At any rate, when the driver rolls the window down, Officer Muldoon catches the strong odor of gasoline. While cars do use gasoline, smelling a strong odor of such in the passenger compartment is not normal. Officer Muldoon instructs the occupants to exit the vehicle and discovers four Molotov cocktails in a box on the floorboards of the back seat. These are considered destructive devices, they are a felony to possess in most states, and you will be called to the scene. You should know you are on solid legal ground to proceed even though no search warrant existed.

Terry v. Ohio

The Constitution is not a suicide pact. That means that anyone's expectation of privacy only goes so far, and there are certain conditions that trump that expectation. One of those is called a "Terry stop." on October 31, 1963, Cleveland Police Detective Martin McFadden observed two men, John Terry and Richard Chilton, acting in what he believed to be a suspicious manner. Both men were walking back and forth in front of some business storefronts stopping to look in the windows and then conferring with each other. After a period of time they were joined by a third man, and Detective McFadden decided the men were acting suspiciously and approached them. He identified himself as a police officer and demanded the men identify themselves. When Terry made what McFadden perceived as a threatening move, he subdued Terry and while doing so he felt a pistol in Terry's pocket. Terry and his friend were arrested and the case eventually worked its way to the Supreme Court as Terry's attorney alleged a violation of his client's Fourth Amendment Rights.

Long ago, the Supreme Court stated that standing up in a crowded theatre and screaming "fire" does not fall within the boundaries of free speech. I would hope that makes sense to everyone, as you can imagine what the results of that might be. I think of Terry in the same light. The Terry stop came from the Supreme Court stating that a law enforcement officer does not violate the Fourth Amendment when he or she stops a suspect and performs a quick search or pat down of the suspect in an attempt to find a weapon. In this case the court decided that the safety and well-being of the officer superseded the expectation of privacy. I think we can all agree that this is a very reasonable decision and that it would be ridiculous to require an officer or fire investigator to allow a suspect to be armed. There are some conditions that must be met here. A police officer cannot just randomly stroll down the street searching anyone he or she chooses. However, the conditions under which you, the law enforcement officer, can perform a Terry stop are much less than probable cause for an arrest. If the officer has a reasonable suspicion that a crime has been committed, is being committed, or about to be committed, and has a reasonable belief that the person may be armed, then a Terry stop is warranted. You will also hear this referred to as a "stop and frisk" or a "Terry frisk."

Administrative Warrants

An **administrative warrant** is neither an arrest warrant nor a criminal search warrant. It is specialized type of warrant that is sometimes used by fire departments for the purpose of inspecting a building for code compliance. The requirement once again comes from the Fourth and Fourteenth Amendments to the Constitution, and two Supreme Court cases established the need for an administrative warrant. The major difference is what is needed to get the warrant. In *Camara v. Municipal Court*, the justices stated that the person seeking the administrative warrant didn't need to provide any probable cause that a code violation had been committed. The applicant only needs to articulate the need for the inspection and the refusal to be inspected. Since fire departments do not seek these warrants on a regular basis, it is a good idea to check with your agency's legal counsel.

Subpoenas

The chances are good to outstanding that if you are involved in any phase of delivering emergency services to the public, you will sooner or later be the subject of a **subpoena**. Subpoenas are generally defined as the power of the court to compel the appearance and therefore the testimony of witness. It is also quite possible that if you are acting as the investigation officer at an arson trial, you may be required to serve a subpoena of behalf of the court. For these two reasons, it is important that you understand subpoenas and the reason they exist. In the United States we have the right to face our accuser and have our day in court. If you happen to be the one accused of a crime, you will be happy that there is a compulsory process to procure both the attendance of your accuser and evidence that might be presented at your trial. That is what subpoenas ensure. In addition to being the subject of a subpoena, you may also be asked to serve them on occasion. It is not uncommon for a deputy district attorney to ask you do this, as you probably know where all of your witnesses are. It would be helpful to know what you are serving.

Subpoena ad Testificandum

A *subpoena ad testificandum* is probably the most common subpoena served, and quite simply, it means that whomever is served with the subpoena is required to appear in court at the appointed time and date and be ready to give oral testimony in some type of proceeding. This type of subpoena is used to compel appearances at several different types of events from actual criminal trials to depositions. The key thing to impart to your witnesses is that the appearance ordered in the subpoena is not voluntary, and there might be severe penalties for failure to appear. The most likely case will be the immediate issuance of an arrest warrant and the eventual arrest of the person failing to comply. In general, in our society we have become increasingly lax concerning authority and consequences in general. This is one area that a command to do something will have consequences.

Subpoena Duces Tecum

A *subpoena duces tecum* is simply a subpoena for the production of evidence. Normally, it will go to the prosecutor from the defense attorney, and you will be expected to fulfill the subpoena by producing every shred of evidence that you have. *Duces tecum* is a term used in the United States and in other parts of the world. It is a Latin phrase that means "bring with you under penalty of punishment." A keystone of our criminal justice system is the right to face one's accusers and know what evidence they have to use against one. There may be reluctance on your part to release every single item in your files, such as your field notes. None of us wants notes that may be scribbled in the field with misspellings and grammatical errors brought to light in court. We would rather our finished report that has been proofread twice after being spell-checked three times be the document of record. It is that, of course, but you will be expected to also produce the notes and sketches you made in the field. The defense is looking for inconsistencies or personal slights toward the defendant that may deflect the specter of guilt from him or her for a while and direct attention toward you. The defense is going to be looking for anything that may convince just one member of the jury that you have a preconceived notion of guilt toward the defendant and that you arrived at that preconception early on in the investigation. You can easily avoid this by taking good notes and always being a professional when you take those notes, write those final reports, and testify in court. Realize that nothing you do from the moment you arrive on scene to investigate the fire is private, and everything you do will be subjected to scrutiny in a court of law.

Key Terms

Administrative warrant. A warrant (as for an administrative search) issued by a judge upon application of an administrative agency.

Arrest warrant. A warrant issued to a law enforcement officer ordering the officer to arrest and bring the person named in the warrant before the court or a magistrate Note: A criminal arrest warrant must be issued based upon probable cause. Not all arrests require an arrest warrant.

Bench warrant. A warrant issued by a judge for the arrest of a person who is in contempt of court or has been indicted.

Exigent. Requiring immediate action or aid, something that is very pressing in nature.

Legal arrest. A seizure or forcible restraint; an exercise of the power to deprive a person of his or her liberty; the taking or keeping of a person in custody by legal authority, especially in response to a criminal charge.

Molotov cocktail. A makeshift bomb made of a breakable container filled with flammable liquid and provided with a wick (usually a rag wick) that is lighted just before being hurled.

Outstanding arrest warrant. An arrest warrant that has not been served. A warrant may be outstanding if the person named in the warrant is intentionally evading law enforcement.

Peace officer. A civil officer (such as a police officer) whose duty it is to preserve the public peace.

Probable cause. A reasonable ground in fact and circumstance for a belief in the existence of certain circumstances (as that an offense has been or is being committed, that a person is guilty of an offense, that a particular search will uncover contraband, that an item to be seized is in a particular place, or that a specific fact or cause of action exists). When supported by probable cause, warrantless search of vehicle may extend to every part of vehicle where objects of search might be concealed.

Search warrant. A warrant authorizing law enforcement officers to conduct a search of a place (as a house or vehicle) or person and usually also to seize evidence; also called search and seizure warrant.

Subpoena. A writ commanding a designated person upon whom it has been served to appear (such as in court or before a congressional committee) under a penalty (as a charge of contempt) for failure to comply.

Review Questions

1. What is the difference between a warrant and a subpoena?

2. What two items must be addressed for a person to grant consent to search a property?

3. What Supreme Court case gives firefighters the right to suppress a hostile fire without a warrant and then allows the fire investigators to conduct an investigation without a warrant or consent?

4. What would be the proper use of an administrative warrant?

5. Under what circumstances can a law enforcement officer enter a property without the benefit of a warrant or consent?

6. If a defense attorney needed to obtain items of evidence from an arson case, what type of a subpoena would he or she need to serve?

7. What Supreme Court decision allows a peace officer to stop a potential suspect on the street and frisk him or her for a weapon?

8. What type of warrant would be issued by a judge if you failed to properly respond to a *subpoena ad testificandum*?

9. Who can issue an arrest warrant?

10. What is the definition of an outstanding arrest warrant?

Discussion Questions

1. Firefighters do not usually have to deal with subpoenas or warrants. As a fire or arson investigator, how might this change?

2. Discuss why *Terry v. Ohio* does not violate the Fourth Amendment to the Constitution.

3. Discuss the importance of maintaining a professional demeanor in the field concerning your notes and sketches.

Activities

1. Talk to a fire or arson investigator and find out if he or she has ever had to apply for an arrest warrant or a search warrant.
2. Talk to a judge or a bailiff and find out how often they have the need to issue a bench warrant.

7

Your Day in Court

Learning Objectives

Upon completion of this chapter, you should be able to:

- Describe the need to dress properly for a court appearance.
- Explain the importance of being prepared to testify at trial.
- Describe the difference of making a mistake during testimony and committing perjury.
- Describe the difference between direct examination and cross-examination.
- Describe the need to maintain a professional attitude while testifying.
- Describe the need to understand and deal with hypothetical questions.

Case Study

Investigator Bob Compton was due to testify in court on a sunny Friday afternoon in an arson trial that had been in progress for about three days. This was the third time that Compton had been scheduled to testify, but circumstances inside the court had prevented him from doing so. As it was a Friday, it was Bob's day off, and he had made plans to attend a party with his wife at a friend's house immediately after his court appearance.

Prosecutor Jackie Gorham contacted Compton by cell phone at about 11 a.m. and told him that he needed to be in court in about five minutes because they needed him to testify before lunch. Compton walked over to the courthouse from Fire Station 1 and took the elevator up to the courtroom for his time on the witness stand. As Compton walked into the courtroom after being called by the bailiff, all eyes turned toward him, and a few giggles went up from the jury box. Compton had chosen to wear an open neck Hawaiian shirt, a pair of white tropical pants, and a pair of sandals into the courtroom. Compton didn't feel like wearing a suit and taking a change of clothing for the party; after all, isn't every Friday casual Friday?

The judge took one look at Compton and told him to approach the bench. Compton did so and was greeted by a lecture in a rather loud voice as to the proper dress for an appearance in court. Compton was then told that he would be scheduled to testify after lunch and that he was expected to present himself at 2 p.m. in proper attire. With all eyes on him, Compton walked out of the courtroom, to prepare for his afternoon now to be spent in court rather than at a party.

The decision to dress casually for a court appearance was a poor decision on Investigator Compton's part. He has now upset a judge in a very important case; he has wasted the time of the prosecutor, the jury, and all the other participants in the case; and he has called his own judgment and credibility into question. This is never a good way to start your day in court.

Introduction

Consider another scenario. It is now about a year to 18 months after the actual fire, and all avenues of plea bargaining and posturing on the parts of all parties have been exhausted. You are finally going to go to trial and ideally put the alleged bad guy away. You are going to be the investigating officer in the case and the star witness for the people.

At this point you need to start considering all of the small items that are going to yield a good performance on your part. It is a performance, and you might as well start thinking of it as such. You are going to be testifying on a subject that you are very familiar with, and you are indeed an expert by the average person's standards. You are going to be telling the truth, the whole truth, and nothing but the truth, so how can it be thought of as a performance? The way you present yourself and the manner in which you dress, speak, and articulate your thoughts are going to have a profound impact on the jury and in all likelihood the verdict they will render. Keep in mind that in all jury trials at least half of the participants are not telling the truth for one reason or another. Some of them just don't know what the truth really is, and some of them will be outright lying about the facts of the case; this is known as **perjury**. We will talk about perjury and how it relates to you a little later in this chapter.

Dress for Success

Let us start with the very basics of testifying in court. The event that you are about to embark on is going to potentially deprive someone of their freedoms and liberties, maybe for the rest of their life. It is a very important and solemn matter and should only be entered into with the greatest respect for the process. With that in mind, consider how you are going to dress to testify in court. This may sound a little silly and trivial, but you are never going to get a second chance to make a good first impression. You are supposed to be the ultimate professional in the courtroom. No one—not the judge, not the prosecutor, not the defense attorney, and not the jury members—know as much as you do about fire behaviors, the crime of arson, or the investigation of this case. It is up to you to dress the part of the professional and the expert in the case (fig. 7–1).

There has been a great movement in our country to "dress down" over the last several years. There was a time when men wore suits and women wore dresses to ride on an airplane; now it is nice just to see someone wearing shoes on a flight. Most businesses have "casual Friday," and some have extended that lack of formality in dress throughout the week. This is not to say these changes are bad things, but a criminal trial is not the place for it. The courtroom is a place where the participants are addressed as your honor, the defendant, the people, and Mr. or Mrs. Smith. The very last thing you want to be is the guy sitting at the prosecutor's table wearing a pair of jeans and a Hawaiian print shirt (fig. 7–2).

Fig. 7–1. You need to dress for success. Your fire department Class A uniform is always appropriate for your day in court, and many juries are impressed with such a uniform.

Fig. 7–2. Resist the temptation to "dress down." Inappropriate dress for a serious situation such as a criminal trial tends to create a disapproving attitude from the jury.

When it comes to courtroom attire, you really have only two choices; a decent suit of clothes or your departmental uniform. Most people like firefighters. Firefighters have a good public image, and the courtroom is a good place to take advantage of that image. Once again, a little common sense has to come into play here. Most fire departments provide their members with dress uniforms that are commonly called **Class A uniforms**. This uniform normally consists of a nicer quality coat, pants, shirt, and tie. The departmental patch is usually on one shoulder, the badge is on the uniform, and, in some departments, ribbons representing awards are worn on the uniform. This uniform is probably worn two or three times a year at special occasions. Testifying in a criminal court could probably be seen as a special occasion, so wearing the dress uniform would be very appropriate. It is also up to you to make sure that other members of the fire department who might be testifying in the case understand the gravity of the situation and dress accordingly.

One problem with court attire is that in some fire departments, the arson investigators are peace officers, and as such they are required to carry a firearm while on duty. While people are used to seeing police officers carry guns even in their dress uniforms, they are not accustomed to seeing firefighters carrying weapons and may not understand the reasons for doing so. In fact, in many fire departments, the chief of the department is not accustomed to the practice, and some chiefs have forbidden investigators from carrying a firearm while in any fire department uniform. This is the managerial rights of the chief, but it leaves the investigator with the decision to either wear civilian clothing or walk around in uniform but unarmed. In many communities and circumstances, it might be a bad decision to walk around unarmed while in uniform.

If you are going to be an arson investigator and if you are going to spend time in court, consider buying the proper attire for testifying: for men, a couple of suits, a couple of shirts, and a few ties; for women, conservative business suits. Consider them an investment in your career and a professional expenditure even if you can't deduct them from your taxes. Try to stay on the conservative side while still exercising your personal taste in clothing. Dark blue, gray, black, and brown are always good and play well with the jury and officers of the court (fig. 7–3). Have the department buy you a badge holder for your belt, and you are pretty much set to go. Once again, if you are required to carry a weapon, it is easily concealable when wearing a suit. You have the option of a shoulder holster, a small-of-the-back holster, or my preferred method, an ankle holster with a smaller firearm.

Fig. 7–3. If your department doesn't have a Class A uniform, a suit and tie or a clean work uniform are both appropriate for your day in court.

Pretrial Preparation

Another major factor in performing well on the witness stand is pretrial preparation. Part of this is going to be your responsibility, and part of it should fall on the prosecutor in the case. Ideally, you will be able to sit down with the prosecutor for an hour or so before the trial and go over the case from the very beginning. In this meeting, the prosecutor should go over all of the questions he or she is going to ask you, and you should go over your answers. There is an old saying that attorneys should never ask questions they don't already know the answer to, this is certainly true in direct examination. The last thing you want to do at this juncture is to trip up the prosecution with an answer they were not expecting. There is no way you are going to be able to know what is going to be asked of you during cross-examination by the defense attorney, but you should plan for unexpected questions by arming yourself with the truth and good set of reports and pictures.

This might be a good time to state that there is absolutely nothing wrong, immoral, or illegal about preparing for your testimony in court with the prosecutor. In fact, if you don't prepare and practice for your appearance on the stand, you are remiss in your responsibilities. The defense attorney is going to try to make a big deal out of you and other witnesses being prepared for testimony by the prosecutor, but it is legal and a very common practice. Almost everything the prosecutor is going to know about your case is going to come directly from the reports you wrote after the fire and possibly from follow-up reports by other investigators and interviews from whatever witnesses you might have. For no other reason that this, you should write your investigation reports with great care and study them before appearing on the stand, just as you would study for a departmental promotional examination or a college exam. In a sense, that is exactly what you are going to be doing once you take the stand—submitting yourself to a series of questions, both friendly and hostile, that will test your knowledge of the fire you investigated. If you studied hard and are familiar with the material, you will do extremely well. But, if you just show up for the test, you will probably not ace the exam.

What Is and Isn't Perjury?

Your name has been called, and it's time for you to take the witness stand. All eyes are on you as you approach the bench, which can be a little bit intimidating. As you start to ascend the steps to where you will be sitting, an officer of the court will stand and ask you to raise your right hand and "swear to tell the truth, the whole truth, and nothing but the truth, so help you God," or words to that effect. You have just been administered the courtroom oath, and by answering in the affirmative you have promised to tell the truth concerning the proceedings in which you are about to participate. To do otherwise could now constitute perjury. Perjury is a serious offense; in fact, it's a felony in virtually every state in the union and a federal crime in a federal trial, and you could and should go to jail for committing this offense. It is important that you know what perjury is and what it is not.

Perjury is the act of willfully lying under oath when the lie you are telling could materially affect the outcome of the trial. For example, if you lie about your age under oath, that is not going to be considered perjury unless your age has some important bearing on the trial. If you testify the fire you investigated was at 4500 Main Street when it was actually at 5400 Main Street, you have not committed perjury, you have transposed the numbers in the address. There is a big difference between making an honest mistake during your testimony and committing perjury (fig. 7–4). The mistake on the address may not enhance your credibility with the members of the jury, but it certainly will not land you in jail either. To commit perjury, your false testimony must be made knowingly and willfully, and it must make a material difference in the outcome of the trial. Those requirements are really what makes it so very difficult to prosecute someone for perjury in a criminal trial, because it is virtually impossible to prove that the false statements made on the stand were made by the witness or defendant with the knowledge that they were false.

Fig. 7–4. When it is your turn to testify, you will be sworn in to tell the truth, the whole truth, and nothing but the truth. Understanding the difference between perjury and making a mistake is important.

Successful prosecution of perjury committed during a criminal trial is indeed a very rare occurrence. In most cases it is impossible to prove that the subject knew that he or she was providing false information or that the information could have changed the outcome of the case. For this reason, many prosecutors will not file perjury charges even though the charges appear to be obvious.

That doesn't mean that arson investigators don't successfully charge and convict several people of perjury; they do, just not from courtroom testimony. In most states when you apply for a driver's license or identification card, you do so swearing all the information you provide is correct under the penalty of perjury, and then you sign that pledge or oath. You would be surprised to learn how many arson suspects are in possession of real state driver's licenses with their pictures on them but with false names and birthdates. It is very easy to file charges of perjury along with the arson filings, and in many states, commercial burglary as well. If you look at the definition of commercial burglary in many states, it is entering a public building for the purpose of committing a felony. The felony in the case cited above was perjury because the person entered the Department of Motor Vehicles with the intent of providing false information to obtain a license. It should become apparent that there are many reasons for completely understanding criminal law if you are going to be an arson or fire investigator.

Direct Examination and Cross-Examination

Back to our court case: You have now been sworn in. You will then be asked to state your name and spell it. After that, you are seated in the witness chair and you are getting ready for **direct examination**. Direct examination consists of friendly questions the prosecutor will ask that you already know are coming. There might be a few spontaneous follow-up questions, but most prosecutors are not going to ask questions that both of you don't already know the answer to. Their job is to present the case on behalf of the people and to prove that the defendant is guilty. Your job is to assist with complete and truthful answers. You can do this with great confidence, because if you didn't know that the suspect was guilty, you would never have arrested him or her and filed criminal charges to begin with. Just remember that knowing someone is guilty and proving it to a jury are two different things.

On direct examination, you want to sit up straight and be attentive to the prosecutor's questions. When the prosecutor is asking the question, have your eyes on the prosecutor and hang on every word. When the prosecutor finishes the question, immediately turn toward the jury, make eye contact, and answer the question in a clear voice loud enough to be easily understandable, but not so loud as to distract the jury. You don't need to stare the jury down, but everyone likes eye contact when speaking with someone or listening to them. It's fine to glance back at the prosecutor, but your answer is for the benefit of the jury.

The reason for answering immediately is twofold. You don't really want to hesitate before answering because it makes you appear confused or unsure of yourself, and it gives the defense attorney sufficient time to think about the question and possibly enter an **objection**, which might stop you from ever answering the question. In many cases you can get your answer in before the defense attorney is able to object, and even if an objection is **sustained**, the jury has heard your answer. In the case of a sustained objection, the judge will probably tell the jury to disregard the answer, but it is very difficult to unring a bell. Once again, there is nothing illegal or unethical about this tactic. You can bet the defense will be using every ploy they can think of to discredit you and your case, and to shift the focus of the case away from the suspect in any manner they can.

When giving your answer during direct examination, you want to make your answers as clear, concise, and complete as you possibly can. Avoid using a lot of technical jargon if you possibly can. You may have to explain how burn patterns led you to discover the point of origin or how a disabled fire suppression system caused you to become suspicious in a warehouse fire. You do not generally need to get into truncated cone patterns and the difference between a pre-action system and a wet system. You can get your point across to the jury without putting on a display of your technical expertise. The level that you connect with the jury might very well make all the difference in the outcome of the trial.

I have had several opportunities to visit with jurors after trials. Most of the time it was after a successful conviction, but there were a few occasions in which we came out on the short end of the verdict. One of the questions I always asked was if there was one single person or factor that caused you to vote the way you did in the jury room. Almost without exception, the cases would turn on the credibility of one or two key witnesses. Remember that the defense attorney doesn't have to prove that his or her client is innocent; all they have to do is to create a "reasonable doubt" in the mind of one juror. That is all it takes to thwart a conviction—a reasonable doubt in the mind of one juror. That doubt can be placed there by an ill-prepared witness who can't convince 12 people that he or she is telling the truth or knows what he or she is talking about. Don't be that witness.

Cross-examination is the exact opposite of direct examination. During cross-examination, you will be questioned by the defense attorney, and you will have absolutely no idea what questions are going to be asked of you. Defense attorneys have only one purpose in the trial, and that is to raise reasonable doubt in the minds of the jury in order to win an **acquittal** for their client or at least cause a **hung jury**. In either case, the defendant is not convicted, and this in itself is a victory for the defense. Our system of jurisprudence is based on all persons accused of a crime receiving the best and most aggressive defense possible. It is important that you as the arson investigator understand this and understand that defense attorneys play by a completely different set of rules. The burden of proof is completely on the prosecution. The defense does not have to prove anything. In fact, the defense doesn't have to put on a defense at all; if they can create enough doubt in the mind of only one juror, they can avoid a conviction. Too many people mistake the term *acquitted* for *proven to be innocent*, but these two terms are completely different. No one in our criminal justice system has to be proven innocent. The defendant is presumed to be innocent until proven guilty beyond a **reasonable doubt**.

Part of establishing this reasonable doubt is to raise questions about your investigation, your qualifications, and your preparation concerning the fire. You can count on the defense to explore every possibility that you violated their client's constitutional rights, or that you violated your fire department's policy during your investigation, or that you conducted a poor investigation and you failed to catch the person actually responsible for the crime. At times, the tactics of the defense may seem to take on a personal nature, but it is important for you to remember that if you were in the position the defendant is in, you would want the same efforts put forth on your behalf.

With that in mind, prepare yourself to answer the defense attorney's questions without appearing to be defensive or hostile to the defense. You want to establish the perception in the jury box that you are there as a disinterested third party who just wants to tell the truth in the case. It is true that you are going to tell the truth and you are hoping to see justice prevail, but by this time in the trial, you are in no way going to be disinterested in the outcome of the trial. Still, if the defense attorney approaches you on the witness stand and says, "Good morning, officer, how are you?" The proper response is: "I am quite well, counselor, how are you today?" By inquiring as to the defense attorney's health, you have actually accomplished two things.

You have made it appear to the jury that you are indeed a pleasant and cooperative person, and you have put the defense attorney in the position of answering your inquiry and maybe put him or her off of their game for just a moment.

Most defense attorneys will ask a little bit about your background and training, but not a great deal, because the prosecution should have established your **bona fides** by the time the defense questions you. I always thought one of the most interesting questions the defense would open up with was something along the lines of: "You don't like my client, do you?" If you can learn to relax on the witness stand, and you are prepared for this question, it can be a great deal of fun and beneficial to your case.

I always answered the question truthfully, and I was always prepared for the follow-up question that you know is going to come right after your answer, and that question is "Why?" If my response was "no" to the initial question, I was prepared with something like, "Well, I am just not fond of people who set apartment buildings on fire with small children inside." If the answer to the initial question was "yes," which oddly enough it sometimes was, the answer might be, "Well, she seemed like a perfectly lovely woman except for the part where she set her boyfriend's Porsche on fire." Here is the great thing about court: if they ask you the question, they have to let you answer. This falls under the category of not asking questions to which you don't want to hear the answer.

Eventually, you are going to be asked the leading accusatory question by the defense attorney. Now is the time to take a leisurely three or four seconds to think about your answer. The reason for this pause is to give the prosecutor time to object to the question (fig. 7–5). Keep in mind that some of the questions the defense is going to ask you are aimed at causing a mistrial or establishing some grounds for appeal. The theory is that when losing badly at a trial, if the defense can somehow cause a mistrial by getting you to say something prejudicial or maybe to get you to mention some past arrest not relevant to the trial, they hope that the district attorney might decide to offer a great deal in a plea bargain to keep from spending the money for a second trial.

There is a really good way to avoid doing this and making yourself the most unpopular guy at the prosecution table. Pause before answering any questions from the defense. Give the prosecution a chance to do their job.

Fig. 7–5. When giving testimony during cross-examination, allow the prosecutor time to object when you are asked a question. Pay careful attention to the instructions of the judge.

There is another way to keep from getting yourself into trouble on the witness stand: Do not offer any information other than what you have been asked by the defense, and do not elaborate on your answer under any circumstances. Remember that these are going to be leading questions posed to you, which is the defense counsel's right. If at all possible you should answer the question either yes or no and leave it at that.

For example, if the defense asks you, "Did you take the samples of the carpet from the fire directly from the fire to the laboratory?" you should say "yes." You should not say, "Yes, because the last time I left samples in the trunk of my car for four days, the sample degraded, and that case was thrown out of court. I learned my lesson!" This is not a good answer even if it is true. It is going to bring up a lot of questions about your department's policy and procedure, and it is going to place a lot of doubt about your competency (and intelligence in my estimation) in the minds of the jury. Remember that the case is about what the defendant did or did not do. Do not assist the defense attorney in shifting the focus to your department's policy and procedure for securing and transporting evidence and samples. The vast majority of the fire departments in the United States do have issues with the collection and preservation of evidence because it is not normally what fire departments do. Larger departments have less trouble than small departments because they have pretty much learned the hard way. Just don't be the person who unwittingly helps the defense. When under cross-examination, keep your answers short—"yes, sir/ma'am" or "no, sir/ma'am"—and, if possible, do not elaborate and give the prosecutor a chance to object before answering. Remember these things, and you will do fine.

Hypothetical Questions

This is an appropriate time to warn you about **hypothetical questions** that will undoubtedly be posed to you while you are on the stand. Beware of these questions. A hypothetical question is one in which the defense attorney makes up a fictitious set of circumstances and then goads you into giving an answer that will make his or her client look innocent. Later, the defense attorney will bring your answer up in his or her closing arguments as if it actually happened, citing your response to the hypothetical question. During the closing arguments, he or she will make a statement such as, "Even Captain Jones stated that it was possible my client might have been framed by the Mafia," neglecting to remind the jury the Mafia question was a ridiculous hypothetical he or she had manufactured for you to answer. I like to call this the "what if the Nazis had Superman" question. These questions usually start out with some type of an opening like, "What would you say if I told you…" and then some fantastic set of circumstances will follow. I realize this sounds ludicrous, but it is a common tool used by defense attorneys in both trials and depositions in attempt to get you, the prosecution's expert, to admit that there is a set of circumstances under which their client is innocent. It is a little bit like an interrogation method used with a lot with arson suspects when a question is posed such as, "What would you say if I told you we found your fingerprints at the crime scene?" Every once in a while that would cause a less experienced criminal to actually admit he or she had been in the crime scene by attempting to come up with an excuse for fingerprints that we didn't actually have. Once again, this is a perfectly legal and ethical question to ask; an innocent person will know that isn't possible and will tell you so.

There is a way to deal with a hypothetical question, and it works every time. Give a hypothetical answer as ridiculous as the set of circumstances that were invented for the question. There is nothing wrong with this as long as you remain respectful and answer in a professional manner. After all, reality was left behind when the hypothetical was posed, and there is no reason for you to inject the real world back into it.

Here is an example of my favorite hypothetical question and answer ever. It actually happened during a trial, and it was a house closer, one of the cleverest answers I have ever heard in court. Two members of our investigations unit had been involved in a fight in a bar on a Friday night. We were in the bar with a couple of police officers to check for overcrowding violations and for valid liquor licenses. As our investigators were attempting to leave the bar, someone threw a beer bottle at them and a fight broke out that eventually ended in the arrest of both patrons and the owners of the bar.

During trial, one of the key questions was who had initiated the brawl. Was it the bottle thrown as the prosecutor alleged, or was the bottle in response to a punch thrown by one of the police officers, as the defense would have the jury believe? As only the investigating officer can remain in the courtroom for the entire trial, the fire department and police department members had not heard each other's testimony, and the defense attempted to take advantage of this. The defense attorney approached the witness stand and asked the following questions, "You claim that you did not throw the first punch during this incident, is that correct?" The officer on the stand answered in the affirmative that he did not throw the first punch. The defense attorney then laid the following hypothetical question on him, "If I told you that your partner stated that you did throw the first punch, who would be the liar?"

So the conundrum here for the officer is obvious, the defense has set him up in a situation where he either has to admit that he is a liar or his partner is a liar. That is not something you want to do, even if the question is a hypothetical one. The answer was brilliant; he looked the defense attorney right in the eye and responded, "Well, counselor, that would be you." The courtroom burst into laughter, including the judge, and the defense attorney returned to the defense table with a stricken look on his face. The witness had just called the defense attorney a liar and made him look foolish to the entire courtroom, and he could not have done it without the help of the defense attorney. Sometimes there is a cosmic justice that supersedes anything we can come up with on our own. That case ended as it should have.

You should also be very cautious when asked multiple part questions, because you can count on the answer to part 3 conflicting with the answer to part 1. It is best to avoid these multiple part questions by insisting on answering them as single questions only. Most prosecutors are aware of this ruse and will object to the multiple part questions. Once again, allow time before beginning any answer for the prosecutor to object, and if the question you are asked does not make any sense to you, state that rather than attempting to come up with an answer. Always remember that when you don't know the answer to a question, it is perfectly acceptable to say so.

Summary

Dress and comport yourself as a professional. Study your reports at length before taking the stand. Once on the witness stand, make eye contact with the party whom you are addressing and speak in a clear and concise way. Listen to the questions from the prosecutor carefully and then answer immediately and with conviction. Listen to the questions even more carefully during cross-examination, and take a substantial pause to think about your answer before starting to speak. During cross-examination, attempt to answer all questions with a simple yes or no when possible, and do not elaborate on your answers. Beware of hypothetical questions, and do not allow yourself to be trapped into saying something you don't really mean to say. Finally, always tell the truth to the best of your knowledge, and if you do realize you made a mistake during your testimony, correct it. Try to relax and enjoy yourself; remember that at the end of the day, no matter how this comes out, you are going home and then back to work the next day. Unless, of course, you didn't read the part about perjury carefully!

Key Terms

Acquittal. Finding a defendant in a criminal case not guilty. The decision to exonerate the defendant may be made by either a jury or a judge after trial. A prosecutor must prove the defendant's guilt beyond a reasonable doubt. A decision to acquit means that the judge or jury had a reasonable doubt as to the defendant's guilt. It may be based on exculpatory evidence or a lack of evidence to prove guilt.

Bona fides. Documentary evidence showing a person's legitimacy; credentials.

Class A uniform. A generally a more formal or dress uniform usually consisting of a coat with matching pants, a dress shirt, and tie. The coat usually has a place to attach a badge and name tag.

Cross-examination. The questioning of a witness at a trial or hearing by the opposing party who called the witness to testify. The purpose of cross-examination is to ascertain the credibility of a witness and to bring out contradictions and improbabilities in his or her earlier testimony, by asking leading questions with the intention of trapping the witness into admissions that weaken his or her testimony. Leading questions are limited to matters covered on direct examination and to credibility issues.

Direct examination. The initial questioning of a witness during a trial or deposition by the attorney who called the witness. It is distinguished from cross-examination, which is conducted by opposing attorneys and redirect examination, in which the witness is again questioned by the original attorney to address his or her testimony on cross-examination.

Hung jury. A slang term for a hopelessly deadlocked jury in a criminal case, in which a decision on guilt or innocence cannot be made. Usually, it means that there is no unanimous verdict, although a couple of states don't require a unanimous verdict to convict. A mistrial will be declared by the judge in the case of a hung jury, and a new trial with a new jury is required. However, the prosecutor can decide not to retry the case.

Hypothetical question. An interrogatory or inquiry propounded to an expert witness, containing a statement of facts that are assumed to have been proven, and requiring the witness to state his or her opinion concerning them.

Objection. In a broad sense, objection refers to an opposition to something. An objection is also a legal procedure protesting an inappropriate question asked of a witness by the opposing attorney, intended to make the trial judge decide if the question can be asked. An objection must have a proper basis, based on one of the specific reasons for not allowing a question. A basis for an objection may include irrelevant, immaterial, incompetent, hearsay, leading, calls for a conclusion, compound question, or lack of foundation.

Perjury. The crime of making a knowingly false statement that bears on the outcome of an official proceeding that is required to be testified to under oath. A statement is made under oath.

Reasonable doubt. It is not a mere possible doubt, because everything relating to human affairs is open to some possible or imaginary doubt. It is that state of the case that, after the entire comparison and consideration of all the evidence, the jury feels that there is a real possibility the suspect may not have committed the crime he or she was accused of.

Sustained objection. Refers to the judge agreeing that an attorney's objection is valid. It usually occurs in the situation where an attorney asks a witness a question, and the opposing lawyer objects, saying that the question is "irrelevant, immaterial, and incompetent"; "leading"; "argumentative"; or some other objection. If the judge sustains the objection, the question cannot be asked or answered. However, if the judge finds the question proper, he or she will "overrule" the objection.

Review Questions

1. What is the difference between committing perjury and simply making a mistake while giving testimony under oath?

2. What is the purpose of taking a few seconds before answering a question while under cross-examination?

3. What is the difference between direct examination and cross-examination?

4. What is the importance of spending time before the trial date reviewing your report and going over the details of the trial with the district attorney?

5. Who should all members of the fire department who are not part of the investigation unit be briefed on before they testify?

6. What are the dangers when answering a hypothetical question?

7. What is a hung jury?

8. Is a suspect being acquitted the same as the suspect being innocent?

9. To what level does the prosecution have to prove a case for a jury to convict a suspect?

10. What is the obligation of the defense attorney to his or her client?

Discussion Questions

1. In the last several years there has been a real tendency in our society to dress down in public. At one time men, women, and children dressed up to attend church, travel, and social functions such as weddings. This is no longer the case, and it is unusual to see anyone in a business suit and tie. Explain why it might be a mistake for a fire investigator to show up to court in casual attire, and discuss what options of dress are appropriate.

2. Discuss the role of the defense attorney in the criminal justice system and why it is important for the investigator to maintain a professional demeanor in the face of cross-examination.

3. Discuss the differences in how you, the investigator, will be questioned by the prosecutor versus the defense attorney.

Activities

1. Interview a local fire or arson investigator and find out if he or she has spent any time in court, and then listen to their opinions as to how they conduct themselves.

2. Attend a trial, and then talk to a jury member after the conclusion of the trial. Ask if the manner in which the police or fire investigators dressed and conducted themselves had an influence on the outcome of the trial.

8

Twenty-Five Miles from Home with a Briefcase

Learning Objectives

Upon completion of this chapter, you should be able to:

- Describe what factors would qualify one to be declared an expert witness.
- Explain why it is important to keep track of the certifications you have earned and classes you have taken.
- Explain why it is important to keep those certifications and documents in a central location that is easy to access.
- Describe the role of expert witnesses employed by the defense.

Case Study

Investigator Bob Compton's Sunday morning started with a call from dispatch, sending him to investigate a car fire and possible assault with a deadly weapon. Compton was told that a suspect was already in custody and had been transported to the jail. He responded and within 10 minutes was on the scene of the fire. Police units on the scene turned control over to him, and he began his investigation. He observed a 2006 BMW 318ti with extensive fire damage to the engine compartment and no signs of forcible entry into the car. The fuel lines appeared to be cut cleanly as if with a sharp instrument, and there appeared to be evidence of flammable liquid patterns on several parts of the engine compartment.

While Compton was conducting his investigation, he was approached by a woman who lived directly across the street. She told Compton that she had seen the car in front of the house several times and thought the woman who owned the house, Lucinda, was dating the owner of the car. She told Compton that she and her husband were sitting on their balcony when they observed an approximately 35-year-old female walk up to the car, unlock the door, open the hood, and then take a large knife from her purse. She further stated they saw the woman cut some kind of line under the hood and then set the car on fire. Compton could not believe his good luck, as he now had two eyewitnesses to the arson.

Six months later at what was supposed to be an open-and-shut case, the defense produced an expert witness who stated under oath that he had conducted tests and research and had determined that this type of automobile had a propensity for catching fire in the engine compartment. He further stated that even though the defendant had indeed cut the fuel lines, he believed the fire could have actually been an accident and the surrounding events coincidences. Compton and the jury sat in a stunned silence while the expert gave his opinion.

Introduction

As title of this chapter implies, everyone knows that anyone farther than 25 miles from home and in possession of a briefcase must be some type of expert. Seriously, an **expert witness** is an individual who, by virtue of his or her education, training, and experience, is deemed to possess specific subject matter knowledge well beyond that of the average person, to the point that he or she can render, officially and legally, an expert opinion during both civil and criminal trials. As we can see by the case study at the beginning of this chapter, not every bit of testimony that is given under the qualification of expert witness turns out to be credible or even slightly believable.

In chapter 7, we discussed your role on the stand as a factual witness, but a lot of the same principles apply to a subject matter expert. The first and most important thing is to get over what is sometimes referred to as the "aw shucks" attitude. The vast majority of us don't like to talk about ourselves in terms of being an expert in anything, and humility is a fine trait in any human being. Unfortunately, modesty is not going to be of much use to the prosecution. The prosecution needs you to be the smartest person in the courtroom, who knows more about fire investigation than anyone else sitting in that courtroom, and who can get qualified as an expert witness for the prosecution.

Direct Examination

You are going to have to help the prosecutor qualify you as an expert witness because he or she probably doesn't know you personally unless you live in a very small town. Even then, it is unlikely that the prosecutor knows about every class you have attended or certificate you have received. It would be outstanding if every fire or arson investigator would take a significant amount of time and develop a really good resume and update that resume on a regular basis. From that resume, you can glean the items that are specific to fire investigation for your **curriculum vitae** (CV). The more detailed your CV is, within reason, the more it will going to help the prosecutor know you. Cases may come up that are somewhat complex and far reaching, and the prosecutor needs to be familiar with your full resume, so have it ready. If you haven't already, start a notebook or scrapbook for original certificates you received for training and seminars. This will be useful for several reasons other than the ones discussed in this book. As your fire service career progresses, at some point you will probably apply for a promotion or an assignment within your fire department, and you may even apply for a job with a different agency. Having these certificates and documentation of classes attended, as well as college transcripts, awards, and items of this type will prove to be invaluable and a real time saver for you (fig. 8–1).

Fig. 8–1. Assisting the district attorney's office in the prosecution of the case is a major function of fire department witnesses. Without your full cooperation, they are at a disadvantage.

Qualifying as an Expert

Now we'll explore what is expected of you and how you will become an expert witness. This is really a simple process: The prosecutor will put you on the witness stand and will ask you a series of questions about your education, training in fire investigation, and your experience with both the fire department and the fire investigation unit. After he or she lays the foundation for you to be declared an expert, the prosecutor will make a motion that the court declare you an expert witness for the purpose of this one specific trial. At this point the defense attorney is welcome to challenge your bona fides or to actually subject you to a series of questions (**voir dire**) to prove your expertise in your particular field. Understand that a true voir dire process is very rarely ever done in the real world, because it would require the defense attorney to actually research and prepare a number of highly technical questions in an attempt to challenge your training and experience. In all likelihood a voir dire is only going to prove that you actually do know what you are talking about and therefore benefit the prosecution. Most of the time, and particularly if the defense attorney has crossed your path in court and knows you, the defense will stipulate to your training and education, and the judge will declare you to be an expert witness (fig. 8–2).

Fig. 8–2. Expect to have your education, experience, and training closely scrutinized as you attempt to qualify as an expert.

The advantage to having expert witnesses on your side during a trial is that in addition to testifying to the facts they actually know about, they can also offer their opinion under oath and for the consideration of the jury. Questions that would otherwise raise an objection from the defense as speculative or without foundation will be allowed in direct examination once you are qualified as an expert witness. You will no longer have to state that you only saw some sheets lying around on the bedroom floor when you conducted your fire investigation; rather, you can state that you saw sheets that appear to have been placed in such a manner as to help the fire communicate from room to room, and that when arranged in this manner, these sheets are called **trailers** (fig. 8–3). You can then go into great detail about what items can be used as trailers and how they are commonly used in arson fires. You might even be able to slide in a picture or two to prove your point! It is easy to see the difference between a strict factual witness and an expert witness in the example just given. Also note that there is a great impact on the jury when the significance of finding the sheets on the floor of the bedroom is explained by the expert witness.

Fig. 8–3. While these may appear to be bed sheets lying in the hallway of a house to the average person, an expert witness will see them as trailers and have the ability to explain that to a jury.

Important Legal Issues

This might be a good time to talk about a few legal issues as they relate to expert testimony. While these legal issues may be complex, their day-to-day impact on us is pretty simple, and it is generally not something we have to worry about. It is. however, something we should know about so we can follow conversations in the courtroom and just outside of it. Knowing what is going on around you is a very good thing.

Let's talk about federal court and federal rules first, even though it is unlikely you will ever wind up in federal court except for a couple of very special sets of circumstances. Federal courts use the **Federal Rules of Evidence** (FRE). They begin with Rule 701, which addresses some circumstances under which a lay or factual witness might provide an opinion in court, and they progress until they actually reach a disclosure that must be filed by an expert witness. In that disclosure the expert witness must reveal not only their training and methods used, but they also must reveal how much they have been paid to render their testimony and opinion. These expert witnesses may never have even examined the fire scene or been involved in your case at all. They will, in some cases, merely review the work of others and, from that review, render an expert opinion for a fee. There is usually a disclaimer in the expert disclosure document that states they are basing their opinion on the work of others and pretty much invoking the garbage-in garbage-out computer theory. Once again, this is not your concern, but it is certainly something good to know about. When you have some spare time, read Rules 701 through 706 of the FRE. It is a good bit of knowledge to have tucked away.

There are a couple of U.S. Supreme Court decisions that you should be familiar with. If you become an arson investigator, you will hear the cases mentioned occasionally, and it is always good to be familiar with current laws and court opinions. Do keep in mind that these opinions might change from time to time, and a year from now, the courts may decide to go in a completely different direction. That is the nature of case law and court decisions.

Daubert v. Merrell Dow Pharmaceuticals (1993) is important to us because it established some guidelines for judges when qualifying expert witnesses. The court case changed virtually nothing for most fire departments and most fire investigators, because they were pretty much already operating under the guidelines *Daubert* set when the case was sent to the Supreme Court. The court's decision defined the trial judge's role as that of a **gatekeeper** with the responsibility of deciding whether an expert's opinion and the underlying investigative methods should be admitted during the trial. The case also set four criteria for the trial judge to consider when admitting or denying expert testimony: whether the method used by the investigator is centered on a testable hypothesis, whether it has been subject to peer review and publication, error rates of the method and professional standards, and general scientific acceptance of the method. In other words, if you state an opinion in court, it is up to the trial judge to decide whether your opinion is based on acceptable scientific methods of investigation or if you are merely pulling these opinions out of thin air.

In 1999 these four criteria were applied in a case called *Kumho Tire Co. v. Carmichael*. The generally accepted result of applying the four criteria established in *Daubert* was that arson and fire investigation must be a science rather than an art form. Once again, most fire and arson investigators never saw fire investigation as anything but a science, and most fire departments nationwide were in compliance with both of these decisions long before they came to pass. They may have changed the way some trial judges made their decisions to qualify someone as an expert witness, but it didn't have any impact on the way most investigators conducted their investigations. For the most part, investigative training and methods had always been based on the four criteria established by *Daubert*, even though most investigators had no idea who *Daubert* was. This should be the case with your department as well.

Responsibilities and Consequences

Being an expert witness also comes with more responsibility and consequences. You will probably remember that earlier in this book it was stated that at least half of all the people involved in court trials are not telling the truth. Some might not realize they are not telling the truth, but some of them are perfectly aware of their perfidious nature. As a rule, expert witnesses who commit perjury are generally dealt with much more severely than a factual witness who gets the facts wrong. There are two very important reasons for this. First, as an expert, you should know what you are talking about, and therefore the intent to deceive is much easier to prove. The second reason is based on the requirement that the falsehoods have, or could have, some material bearing on the outcome of the trial. Once again, if you are an expert witness for either side, you are there with the expressed intent of having a material influence on the outcome of the trial. Keep in mind that being an expert witness means that you have education, training, and experience in your particular field that exceeds that of the average person and which caused the judge to declare you an expert. It does not necessarily mean that you know everything in the world about anything that you might be asked. If you don't know the answer to the question or you don't have an opinion concerning a question, say so. You are generally not going to be asked any surprise questions by the prosecution when you are an expert for the same reason you didn't get asked any surprise questions by the prosecutor when you were a factual witness. You are going to sit down with the prosecutor before you ever take the stand, and you are going to know what you will be asked, and you should know how you are going to answer.

Keep in mind that testifying as a factual witness and an expert witness can be done simultaneously without the prosecutor declaring that he or she is asking an opinion question. Once you are qualified as an expert, the line will blur, and the questions will be both factual as to your investigation and expert in the conclusions you have drawn from your investigation.

Cross Examination

Once you are finished with direct examination, you will be subjected to cross-examination as both an expert witness and a factual witness. As in direct examination, the line blurs as to what role you are playing. Although the defense attorney will not be questioning your qualifications as an expert, he or she will most definitely be disputing your conclusions if they have the potential to harm their client. It is very likely you will be required to defend your credibility, and this is the most common place that you are going to be subjected to hypothetical questions in which you are asked to draw a different conclusion that will run counter to your expert opinion and make it appear that you may have deed drawn an improper conclusion. From the example earlier in this chapter concerning bed sheets that were used as a trailer to spread the fire from room to room, a typical hypothetical question might be, "Is there any time in your life that you have seen bed sheets lying on the floor and they were not being used to cause a fire to spread from room to room?" Of course, we have all seen bed sheets lying on the floor waiting to be laundered or put down to keep a pet from lying on a bare floor. The trick here is for you, the expert witness, to cite training and previous experiences that allow you to differentiate the situations in which you have found bed sheets on the floor. If you have been an investigator long enough to be qualified as an expert witness and you have the education and certification to be declared an expert witness, you should certainly be able to answer that question.

You Might Not Be the Only Expert in the Room

You should also not be surprised if the defense produces an expert witness who directly disputes your findings (fig. 8–4). The key thing to remember here is not to take this personally or in any way let it sway your opinion and testimony. The fact of the matter is that in some cases honest and well-informed professionals disagree on scientific matters. The defense witness's experience may differ from you, and his or her opinions may be based on that experience, just as your opinions are based on your experience. There is also the possibility that he or she is willing to say whatever the defense needs him or her to say to benefit the defendant. While you might find this repugnant, you must understand that this is a common practice, and one that if you spend any time at all in court, you will encounter. When faced with what you consider to be an unscrupulous defense expert, maintain your professional demeanor and testify truthfully and without malice. There is nothing you can do to control the actions of others, but you can control the manner in which you conduct yourself. Never lose that professional edge.

Fig. 8–4. Do not expect to be the only expert in the room. The defense will have the right to produce its own expert to dispute your theories or opinions.

You are also going to be amazed at how many different fields in which you might be qualified as an expert. Most fire investigators started as firefighters and have a number of areas of expertise. It is not unusual for fire or arson investigators to have at one time in their careers been paramedics or hazardous materials technicians. It is quite possible that you may serve as an expert witness in more than one subject. That is really up to the prosecutor and judge to decide if you are qualified to do so.

Case History II

At one trial I made the statement that I believed the suspect I arrested was intoxicated when he set his girlfriend's house on fire. Voluntary intoxication is not a defense for a criminal act in most states, and it actually does help explain how inhibitions were lowered to the point of committing a felony when provoked. My opinion drew an immediate objection from the defense attorney who wanted to know my medical background for recognizing someone who is intoxicated.

The judge asked me to comply, and so I ran down my paramedic training at UCLA Medical Center, seven years as a mobile intensive care unit paramedic, and my paramedic preceptor training, and then how I had help train, as a paramedic preceptor, five paramedics for both my department and surrounding fire departments. The judge then asked how many intoxicated individuals I might have encountered in those seven years, and I honestly answered I believed it to be hundreds. Based on my education, training, and experience, I was declared an instant expert on recognizing an intoxicated individual, and my testimony was allowed.

You may have been an electrician before you became a firefighter, or maybe you were an armorer in the Marine Corps. If either of these things is true, you may possess specialized and expert knowledge about electrical wiring or firearms that may prove useful someday, so make sure it's in your CV.

Summary

Qualifying and giving expert testimony bring with it a responsibility to testify to the best of your knowledge and ability. Being an expert allows you to introduce your opinion into a criminal trial based on your education, training, expertise, and experience as perceived by the judge in the case. Expert witnesses may also give factual testimony during the trial, and the two are sometimes intermingled during questioning. The defense has every right to question not only your factual testimony but also your expert testimony, and indeed they may decide to produce their own expert whose opinion may very well differ from your opinion. Learn to accept these differences as part of the legal process and to conduct yourself in a professional manner at all times.

Key Terms

Curriculum vitae (CV). A short, written document that provides an overview of a person's experience, education, and other qualifications. It is generally more condensed and specialized than a resume.

Expert witness. A witness who has knowledge beyond that of the ordinary layperson enabling him or her to give testimony regarding an issue that requires expertise to understand. Experts are allowed to give opinion testimony, whereas a nonexpert witness is not allowed to do so. In court, the party offering the expert must lay a foundation for the expert's testimony.

Federal Rules of Evidence (FRE). The code of evidence law governing the admission of facts in the U.S. federal court system. FRE is applicable to both civil and criminal trials.

Gatekeeper. A gatekeeper is a person who has the authority or ability to control access to a decision maker or to certain information. In the context of legal proceedings, the judge in the case occupies the position of gatekeeper during the trial.

Trailers. In reference to fire investigation, trailers are sheets, towels, or any material that is fashioned in a manner to create "trails" from one room or area to another room or area to aid in the spread of fire.

Voir dire. A Latin term meaning "to see or speak." Voir dire is a legal procedure conducted before trial in which the attorneys and the judge question prospective jurors to determine if any juror is biased and/or cannot deal with the issues fairly, or if there is cause not to allow a juror to serve. Voir dire is also employed to reveal the justification to confirm or deny a motion to have a witness declared an expert.

Review Questions

1. What items are considered when a prosecutor or defense attorney is asking the judge to declare someone to be an expert witness?
2. What is the difference between an expert witness and a factual witness?
3. What is a curriculum vitae (CV)?
4. What does the term voir dire actually mean?
5. Who occupies the role of gatekeeper in deciding who will be allowed to act as an expert witness in any case?

Discussion Questions

1. Discuss under what conditions two individuals both qualified in court as expert witnesses in the same field might differ in testimony.
2. Discuss what requirements *Kumho Tire Co. v. Carmichael* placed on fire investigators and how they might affect testimony.

Activities

1. Gather all the certificates that you have accumulated and create a CV that would be useful to an attorney in qualifying you to be an expert witness.

2. Attend a trial in which expert testimony is going to be considered, and watch the voir dire process and the testimony of the expert.

9

The Aftermath

Learning Objectives

Upon completion of this chapter, you should be able to:
- Describe the impact of the crime of arson on family members and articulate the need to keep them informed during the trial process.
- Describe the need to keep all parties involved aware of changes and the progression of the trial from the chief of the department to the public information officer.
- Describe the value of speaking with jurors after the end of the trial and learning the reasons for their decision.

Case Study

After the verdict is in and read, I think that there are still some obligations on the part of the investigator that should be fulfilled. Don't forget that in every case there is at least one victim, and many times there are several victims. People who have lost their homes or property and, in some cases, loved ones to the arsonist you have just convicted deserve to be among the first to know of both good and bad news. Unfortunately, they are often completely forgotten.

The last case in which I was involved in Long Beach was a murder by arson trial that had been going on for several years. The case did not actually come to trial until well after my retirement. The lead investigator was a captain I had trained and who took my place after I retired. I was needed at the trial, and it was necessary for me to travel from Oregon to Southern California a few times to testify. I had arrested and convicted the defendant a few years earlier for the same type of an arson crime, but in that case no one had been killed or injured. During the time that I spent back at my old haunts, I had occasion to meet the family of the victim and to get to know them in a very small way.

At the end of the trial and immediately after the verdict, we drove to the home of the children and other relatives who had lost their father in the fire. This investigator, my replacement, who had spent a phenomenal amount of time and effort working this case, broke the news of the conviction to the family, thanked them for all of their cooperation, and once again consoled them for their loss. The reaction of the family was dramatic and was worth all of the time he and I had put in on this case and all the time I spent traveling back and forth from Oregon to Long Beach. It was easily one of the most gratifying moments of my career, and one that I wish everyone could experience. I cannot begin to tell you how proud I was of him and my fire department. For at least this one moment in time, we had prevailed and had helped win justice for the family of the victim in this horrific crime.

Introduction

When the trial is finished, the jury will retire and deliberate the fate of your case and that of the defendant. One of the things that many fire investigators learn early on is that there is absolutely no way anyone can draw any conclusions from the look on the jurors' faces, the day of the week, or how long they deliberate the case. Many investigators will tell you that when a jury comes back quickly with its verdict, the verdict will usually be guilty. The simple fact is, when you investigate a case properly, write good reports, and prepare yourself for trial, and the prosecutor puts on a good presentation, you are going to win your case. Most fire investigators will probably tell you that they never presented a single investigation to the district attorney's office for prosecution in which they were not 100% convinced that the defendant was guilty. You will probably never be able to find an instance of a deputy district attorney (DDA) filing a case in which he or she did not examine the evidence carefully and come to the same conclusion. Successful prosecutions are only achieved through the hard work and the cooperation of a number of people besides the obvious ones.

Many fire investigators will also tell you that the cases they were involved in where an acquittal was won by the defense were usually the result of a major mistake the prosecution made during the trial, a poor witness, or a mistake on the part of the fire department. This may run contrary to the public's perception of the fire service and our perceptions of ourselves, but we do make mistakes and fire investigations are one of the areas where we make them the most. There are several reasons for these mistakes, and some of the time we can do something about it and some of the time we can't. Fire investigators realize that evidence will get moved around during fire suppression operations by hose streams and firefighters moving through burning structures during rescue and extinguishment. There is nothing that can be done about that; we need to extinguish the fires and save the people in the burning buildings. Most fire and arson investigators began their careers as firefighters, and they fully understand and accept the importance of these operations and the havoc they can wreak on what is eventually going to be a crime scene. No one would ever suggest that we should alter our firefighting strategies and tactics in an attempt to preserve a crime scene. That would be a frivolous and unacceptable suggestion.

However, there is also a good deal of evidence destroyed or spoiled beyond the point of value during overhaul operations, and this can and should be avoided. The plain truth is some firefighters don't know any better and need to be educated, while others just don't care. Fire and arson investigators are generally not particularly popular in the fire department hierarchy for a number of reasons, which we'll talk about next.

Whose Job Is It Anyway?

A lot of firefighters, from the chief engineer down, don't feel that fire and arson investigation should be done by the fire department. They view it as a law enforcement function and feel that the police should be handling the job instead of firefighters. They see it as something that just generates more work for them, takes up valuable monetary and personnel resources, and in the end creates citizen complaints when someone is brought to answer for their unlawful actions. Others view fire investigation as tying up fire engines and trucks and keeping them on a fire scene longer than is absolutely necessary.

While there may be some validity to both arguments, most fire department investigators believe that it is a fire department responsibility, and it is clearly stated to be so in the International Fire Code.

> Section 104.10 Fire Investigation (2012 IFC): The fire code official, the fire department or other responsible authority shall have the authority to investigate the cause, origin and circumstances of any fire, explosion or other hazardous condition. Information that could be related to trade secrets or processes shall not be made part of the public record except as directed by a court of law.

In all likelihood whatever hybrid version of the International Fire Code your department has adopted will also place the responsibility with the fire department. My point is that if you have decided to do fire investigations, then do them. If you don't want to do fire investigations properly, then get out and cede the responsibility to another agency. What the fire code does not cover is to what degree the fire department accepts and discharges this responsibility, and debates rage as to what level the fire department should be involved. Doing the job halfway or involving three or four different agencies rarely works out well and usually does not produce a successful end to the investigation. Often we forget that our responsibility is to the victims and those otherwise affected by arson, and we are much more concerned about territory and perception.

The Verdict Is In

After the jury returns with its verdict, the prosecutor and defense team will be summoned to the court to hear the verdict. If you are lucky, you are involved with a prosecutor who cares enough to give you a call so you can be in court to hear the results of your hard work. The relationship that you enjoy with the district attorney's (DA's) office will vary from DDA to DDA and from investigator to investigator. To me, missing the reading of the verdict was always a little bit like leaving a baseball game in the top of the ninth inning with the score tied. You're going to find out who won later, but it isn't the same as being there to see the winning run scored.

Talk to the Members of the Jury

Not being there will also lessen your chances of talking with the jurors after the case is over, and there is a lot of knowledge to be gleaned from this exercise. Once the case is over and the jurors are dismissed, they are free to talk to you and anyone else about any aspect of the case (fig. 9–1). They are under no obligation to talk with you, but they are free to do so. Most jurors are quite willing to chat after the case, because they usually have many questions they would like answered. You will, of course, encounter the occasional juror who wants nothing to do with either side and just wishes to be left alone after the trial has reached its conclusion. This juror has every right not to speak with you, and you should respect that right. Imposing on this juror will probably not be a pleasant experience, and any information you do gain will be of dubious value. This talk that you are going to have with the jurors is a give-and-take that could greatly benefit you in the future, and you should make every effort to speak with as many of them as possible after the trial is over, win or lose. A lot of investigators don't take the time to seek out jurors and talk with them, and a lot of investigators just don't get the opportunity to do so or don't realize they have the right to talk with cooperative jurors. If you don't attempt to interview these fine citizens who served on your jury, you are missing an excellent opportunity to make yourself a better investigator. The information you obtain from them might be critical in your next case.

Fig. 9–1. If the members of the jury are willing to talk to you about the verdict and trial, you should do so. Use this information to make you a better investigator in future trials.

What Kind of Information Can Be Gleaned?

It is always important to learn what witnesses the jurors found compelling and what witnesses they dismissed. It is enlightening to know what pieces of evidence they found convincing and what they found to be confusing and how it could have been presented in a way that made more sense. You would certainly want to know if there was one single piece of information that caused them to cast their ballot for conviction or acquittal. The information that is available to you if you take the time and make the effort may be absolutely critical to the success or failure of the next trial you participate in. From the standpoint of learning—and becoming a better fire investigator and more effective at trial—it really doesn't matter whether you won the case or lost the case. In fact, you can probably learn a lot more from your failures than your successes.

Most of the questions the jurors will have for you will deal with things they wondered about during the trial but couldn't be brought up for fear of causing a mistrial. They are almost universally curious about the defendant's arrest record and criminal history. Sometimes they are interested in your investigative techniques, and sometimes they just want to talk about your job and how interesting they found it to be. Remember this is a two-way conversation, and until you walk away from the discussion, there is always something to be gained. It is important for you as the investigator to remember that jurors are citizens who have given up a few days or a few weeks of their lives to serve. As a public servant, you owe them the respect and the courtesy of answering their questions if you can.

Visit with the Victims Immediately after the Trial

Many fire investigators believe that their job is finished as soon as there is a verdict in the trial; this is just not the case. It is really important for all of us to understand that for every criminal we are able to bring to justice, there is at least one victim, and in some cases, there may be more than one victim. While we go home at the end of the day to our families and forget about work for a little while, many of these victims live with the results of the crime 24 hours a day. They live in fear of retribution from the friends and family of the arsonists, they live in fear of an unfavorable verdict at trial, but most of all, many of them fear that their contact with you will end immediately, and they will never achieve closure. Many of them feel this way because their other experiences with law enforcement have ended that way.

It is important that you remain in contact with the victims on a regular basis all the way through the trial and then in some case for a significant period of time after the verdict is in (fig. 9–2). In many jurisdictions the sentencing portion of the trial does not occur until several weeks or months after the verdict, and the victims of the crime have fears the suspect will now have nothing to lose and come after them. In some of these cases, their fears are justified. It is important that you are honest and inform them of all developments in their trial, both good and bad. Most people can accept bad news, but most people do not respond well when there is no word at all. You have asked for their cooperation for the last several months or even years, and a little cooperation in the other direction will go a long way.

Fig. 9–2. Pay a visit with the victims and witnesses after the trial. Remember that they invested a good deal of their lives in your case and deserve to know what happened.

Visit the Fire Crews

Don't forget to pay a visit to the crew who fought the fire and led to your successful investigation. Don't forget to tell them how their efforts to preserve the evidence and to point out important information to you during the fire helped you make this important case. Don't forget to tell them how important they were and how you couldn't have convicted the arsonist without them. Everyone likes to hear about when they do a good job. As human beings, we respond well to a pat on the back or a kind word from our peers. Praise from you might cause them to put an even better effort into the next fire.

Don't forget to visit the fire chief and tell him or her about the great case you, the fire department, the police department, and the DA's office were able to prosecute by working together toward a common goal. Don't forget to thank the chief for all his or her support in your efforts to bring an arsonist to justice. The fire chief is generally in charge of the entire fire department, including budgetary considerations down the line. It is always a great idea to let those who control the budget know that the taxpayers' money has been well spent and that you stand ready to do a great job at the next fire investigation, which could occur at any minute.

However, you must understand that the job you do is probably going to bring a lot more in the way of complaints than it is in the way of compliments and praise. If you are looking for everyone to like you and to be glad to see you, then this is the wrong job. This especially true if you are in the position of being a peace officer with the power to make arrests. The concept of arresting people and depriving them of their freedom is foreign to most firefighters, including the chief of the department. For every telephone call that the chief gets from a grateful citizen, there will be 10 from the family and friends of the arsonist claiming that you have framed their family member and you are doing your best to put a completely innocent person in prison. The chief is not going to like this and may start to view the fire investigation, and you, as more trouble than you are worth. So when you have a case where everything went well and you put the bad guy in prison and protected the people, you need to let the chief know.

The People Have a Right to Know

Don't forget to visit with your public information officer and tell him or her about the successful conclusion of this case and the protection it provided for the people of your community. Fire investigation is not something that is generally done in full view of the public, and in fact, many people don't really understand that many fire departments conduct investigations at all. Many times the reporters who write stories about arson investigations or stories about arsonists being convicted assume that this type of investigation is done exclusively by police departments. While the police department deserves full credit for the good things they do, so do we. Public information officers love to write positive articles about fire departments and get those articles published in the local newspaper; good news and good public relations make for a happy and appreciative public. What public information officers and reporters like even more than that is when someone else writes a really good story about a fire investigation and then hands it over to them for publication. It would behoove you to develop a relationship with a few reporters who work for the local newspaper, radio, or television stations. It might cost you a lunch or two, but when you need or want a story to appear in print or on television, you might find it easier to accomplish. Consider doing all these things, and you might find an increased level of cooperation and involvement in your next fire investigation.

Summary

The entire point of having a fire investigation unit is to discover arson fires, find the person or persons responsible for those fires, and then bring them to justice. This isn't done by one or two people. A successful prosecution is the result of cooperation among many people representing many agencies. When we are successful, that success should be shared with the people who contributed and with those who support our mission.

Ultimately, the victims, taxpayers, and residents of the community have every right to know about the hard work that went into the prosecution. Unless we provide that information to the appropriate news sources, they never will.

Review Questions

1. Why is it important to spend time after the trial talking with jury members and hearing their opinions about the trial?

2. Where do fire departments draw their authority to conduct fire investigations?

3. Why is it important to speak with the victims of the crime after the trial?

4. Why would it be important to talk with the fire crews who participated in extinguishing the fire you investigated?

5. What role could the public information officer in a fire department play in relation to fire investigation?

Discussion Questions

1. Why is fire investigation generally regarded to be a fire department function? What is the basis of your answer?
2. Why might it be a good idea to brief the chief of the department on the outcomes of your arson trials?

Activities

1. Talk with some fire or arson investigators in your community who have been to trial in an arson case and find out how much time they spent with the victims of the crime both before and after the trial.
2. Talk with the chief of the department and the public information officer and find out what kind of interest they have in fire investigation and if they view it as a fire department responsibility or a police department responsibility.

Index

U

V

W